HYPNOSIS
COMPLICATIONS

HYPNOSIS COMPLICATIONS

Prevention and Risk Management

By

FRANK J. MACHOVEC, PH.D., A.B.P.H.

AN AUTHORS GUILD BACKINPRINT.COM EDITION

iUniverse, Inc.
Bloomington

HYPNOSIS COMPLICATIONS
Prevention and Risk Management

AN AUTHORS GUILD BACKINPRINT.COM EDITION

Published by iUniverse, Inc.

For information address:
iUniverse
1663 Liberty Drive
Bloomington, IN 47403
www.iuniverse.com
1-800-Authors (1-800-288-4677)

Originally published by Charles C Thomas • Publisher

Because of the dynamic nature of the Internet, any Web addresses or links contained in this book may have changed since publication and may no longer be valid.

The views expressed in this work are solely those of the author and do not necessarily reflect the views of the publisher, and the publisher hereby disclaims any responsibility for them.

Any people depicted in stock imagery provided by Thinkstock are models, and such images are being used for illustrative purposes only. Certain stock imagery © Thinkstock.

ISBN: 978-1-4759-6003-7 (sc)

Printed in the United States of America

iUniverse rev. date: 12/04/2012

ABOUT THE AUTHOR

Frank MacHovec has B.A., M.A. and Ph.D. degrees in psychology and is a clinical psychologist licensed in Virginia, North Carolina and Manitoba, Canada. He received his initial training and experience in hypnosis in 1970 from Michael M. Miller, M. D., author of *Therapeutic Hypnosis*, then 1973-1975 with F. L. Marcuse, author of *Hypnosis: Fact and Fiction* and *Hypnosis Throughout the World*. He was Chairman of the Ethics Committee, Vice President, then President of the Manitoba Hypnosis Society. He has attended advanced seminars in clinical hypnosis in 1978, 1979 and 1981 at the Southern California Society for Clinical Hypnosis and served as faculty at an annual conference of that society. He has taught the Psychology of Hypnosis at the University of Lethbridge in Alberta Canada.

He has served as Chairman, Secretary or as a member of ethics committees of professional associations in Manitoba, Canada, and in Alaska and Virginia. In 1981 he was awarded a Diplomate in Clinical Hypnosis by the American Board of Psychological Hypnosis. He is a member of the Society for Clinical and Experimental Hypnosis and the International Society of Hypnosis. In 1982 he was awarded a National Certificate of Recognition by the Division of Psychologists in Public Service of the American Psychological Association for his work in establishing ethical standards for hypnosis in Alaska. He has presented at state, regional, national and international conferences on the need for more care and higher ethical standards in the use of hypnosis.

In 1984 he chaired a committee to draft proposed restrictive legislation to control the practice of hypnosis in Virginia. The same year he founded the Center for the Study of the Self, a nonprofit clearinghouse for research and case material on hypnosis complications available to all the professions. In 1985 he presented on hypnosis complications at the 10th International Congress on Hypnosis and Psychosomatic Medicine in Toronto, Canada. The response to his talk there was so favorable that he further researched the subject, adding to his original material, resulting in the present volume.

*". . . as with any treatment,
do no harm to the patient."*

Hippocrates

To Evelyn, for her patience (and proofreading!), Jacquie and Frank for putting me up (and putting up with me!) and to Allison who re-educates us all about life and love.

ACKNOWLEDGMENT

My thanks to the late Michael M. Miller for hours of instruction and supervision, Andrew Salter for precious moments of inspiration and sound advice, Fred Marcuse for his gentle warmth and humor, and to Michael J. Rostafinski, M.D., consultant, colleague and friend for "IV feeding" (information validation) in the final hours of this endeavor.

FOREWORD

IN 1970, when I first learned of hypnosis from Michael Miller, M.D., careful and methodical graduate of the University of Vienna, himself taught by Freud, there was little concern about complications. Hypnosis was safe. During the next sixteen years, in my own practice, at conferences (especially chats before and after formal sessions), in case consultation, teaching, and on three ethics committees in the U. S. and Canada, I learned of many "horror stories," of unexpected problems from a few moments of minor discomfort to some really serious medical and psychiatric emergencies which persisted for days and weeks. Hypnosis is not as safe as many of us have been taught. Complications are under-reported and underestimated in type and frequency.

This book is written for clinicians and researchers and for interested consumers who want to know more about hypnosis and adequate standards of care for its use. Research methodology and theoretical issues are avoided except as they relate to complications. The goal was to provide a convenient, practical source of information on risk management and preventive practices to avoid the problems here, taken from several decades of books and journal articles and the author's own personal and professional journey from false security to cautious awareness.

It is expected some will criticize this work for its lack of detail in case descriptions, too few statistical analyses, and reliance on anecdotal records. These are justified criticisms! They point up the need for more data. We need more detail about what can go wrong. A suggested complications report form is included as an appendix to this book. The lack of data made it difficult to evaluate risk factors. Sixteen years of experience using hypnosis was a great help. Adding to this difficulty is the unfortunate fact that most published theorists and clinicians report few failures. There is little research based on large numbers of subjects and what there is relies heavily on college student volunteers, usually psychology majors on campus, not clinically "the real world."

Clinicians may be concerned, perhaps defensive, to read of potential danger from techniques they use routinely (and without apparent ill effect) and which they have been assured are "safe." They may believe, like combat soldiers, "it won't ever happen to me." Regrettably, it does happen to someone. They may feel unwanted side effects would happen anyway even without hypnosis. It is sometimes difficult to separate "what belongs to hypnosis and what belongs to therapy" (J. Hilgard, 1974, p. 282). The unhappy fact is that every problem cited here coincided with the use of hypnosis and did not occur at any other time. The vast majority of casualties had no prior history of referral for or treatment of mental problems. The conclusion is inescapable: there was something about hypnosis or the subject's mental state before and during hypnosis that precipitated complications (possibly both).

Some may feel that dwelling only on complicatons is misleading and exaggerates what are really minimal risk factors. Few successes are reported here because of the book's focus. I have used hypnosis sixteen years and intend to continue to do so, but not as I was taught it. Automobile and air travel are safe despite the accident rate. So also for hypnosis, except there is no accident prevention program for hypnosis and there should be. This book is an attempt to provide the basis for an effective program of risk management, to help clinicians and researchers integrate preventive practices into their work as standard practice. It is an appeal to all who use hypnosis to be more aware of potential problems, to be more careful and more caring.

Hypnosis has helped alleviate mental and physical suffering for centuries and there is no doubt it will continue to do so. It has helped us better understand the nature of personality, consciousness and the variety of mental processes. It has done so because it is a powerful tool — and, because it is, it should be carefully used, with the same skill and the same educational and training requirements as any other specialized scientific method or technique. Hypnosis can be a laser of the mind, and its use must be restricted to protect the public.

Finally, the incidence of complications has diminished in my own practice by using the preventive practices described here but mostly by being aware far enough in advance to avoid the pitfalls. May it be so for you!

Frank MacHovec, Ph.D., A.B.P.H.
Center for the Study of the Self
3804 Hawthorne Avenue
Richmond, VA 23222

CONTENTS

HYPNOSIS
COMPLICATIONS

CHAPTER 1

HYPNOSIS: ITS PRACTICE, ITS PROBLEMS

It is no easy task
to pick one's way
from truth to truth
through besetting errors

—Peter Mere Latham
(1789-1875)

HISTORICAL ANTECEDENTS

CASE 1. She was a typical teenager to all who knew her, of average intelligence and normal psychosocial development never referred for any mental problems, quiet, somewhat introverted, a bit passive-aggressive but who got along well with family and friends and in school. It was not at all unusual for her to join a few friends to see a "hypnosis show." Within hours she was mute and stuperous, unable to speak, recognize anyone or anything, eat or go to the bathroom. Her tongue slid backward into her throat and she gasped for air, strangling, eyes rolled up in their sockets. Hospitalized, she lay helpless for a week, fed intravenously. It took seven days of intensive medical care just to keep her alive, three months of psychotherapy to return her to some semblance of herself and to prevent relapse (Kleinhauz and Beran, 1981, pp. 148-161).

There was only one event separating the few hours of entertainment at the "hypnosis show" with her friends and lying helpless, mute and choking on her own tongue. She had been hypnotized. Admittedly, this is a dramatic case which lends itself to sensationalism, a "horror story" likely to evoke sympathy and great concern from those who read it. Fortunately few hypnotists will ever encounter such a case in all the years of

3

their practice. Unfortunately it did occur in someone's practice and the use and role of hypnosis in precipitating crisis and treatment of it are well documented (Kleinhauz & Beran, 1981, pp. 150-155). There are other "horror stories" of problems arising from or coincident with the use of hypnosis, from earliest sources 2000 years ago to the most recent professional journals.

CASE 2 was Hermo of Pasos who was successfully treated "for blindness" at a "temple of sleep" of the Cult of Asklipios in ancient Greece. He did not pay the required fee, however, and he was blinded again. He returned to the temple, this time paid the fee, was again successfully treated, and did not need to return (MacHovec, 1979, p. 87).

It is not clear whether Hermo was totally without sight or suffered from an organic or functional, hysterical blindness. The latter is more likely in view of the relative ease the symptom was removed and replaced. It is also not certain hypnosis as we know it today was used 2000 years ago in Greece (Stam & Spanos, 1982). Temple statuary depicted the god Asklipios, "the most humane of all the gods" seated and staring a lion into submission, or at least with steady eye contact with a lion in a passive, seated position. The inscription reads: "The god who disperses sleep from his eyes" (Edelstein & Edelstein, 1945). The imagery, symbolism and phraseology and routine use of pallets where patients slept "to hear the god speak to them" suggest hypnosis or some similar phenomenon (MacHovec, 1979).

Shakespeare wrote *The Tempest* at the end of his career, after *MacBeth* and *Hamlet,* plays rich with descriptions of auditory and visual hallucinations, paranoia, major depression, and schizophrenia, according to current diagnostic classification. To write such descriptions, the author had to have a very good understanding of mental processes. It is interesting that *The Tempest* was written when Shakespeare was at his peak, at full maturity as a playwright, producer and director. *Tempest* describes and contrasts two kinds of altered state, one with a physical cause, alcohol, and the other caused by a mental process, strong suggestion, persuasion or telepathy (MacHovec, 1981). The resultant behaviors were generally unexpected and unwanted by those who experienced them.

In the 1780s, a hundred years after Shakespeare described altered and disturbed states of mind, Franz Anton Mesmer (1734-1815) "induced animal magnetism" initially by the "baquet," a wooden tub of water with iron rods protruding from it. Patients were "mesmerized" by touching or holding one of the iron rods. Some of the more susceptible subjects were entranced, simply being in the room or near Mesmer himself.

CASE 3 was a man who was "mesmerized" in 1784 by holding a baquet rod, and it was reported he "felt considerable heat, first in the pit of his stomach, then through his whole body . . . followed by nausea and an urge to vomit which he could avoid only by abandoning the iron rod of the baquet" (Shor & Orne, 1965, p. 13).

This case was reported to the Royal Investigating Commission convened by order of King Louis XVI who appointed the American ambassador, Benjamin Franklin, its "president." A supplemental report of this commission in 1784 "stressed possible dangers, observing that some patients who showed no symptoms before being mesmerized experienced symptoms afterwards" (Kaufman, 1962, p. 895).

With the natural curiosity for which he became famous, Franklin conducted his own investigation in which seven subjects were mesmerized (Shor & Orne, 1965). In what may have been the first experimental investigation of hypnosis complications, three of the seven subjects experienced side effects: two developed headaches "when touched by the mesmerist" and one complained his eyes "hurt and watered." Franklin's "casualty rate" of unwanted after effects was 42.8%.

Sigmund Freud used hypnosis extensively in the earlier years of his practice, a hundred years after Mesmer. Freud visited Charcot and Bernheim in France and was quite knowledgeable in hypnosis theory and techniques of trance induction. He discontinued its use, explaining: "When I attempted to apply to a comparatively large number of patients Breuer's method of treating hysterical symptoms by an investigation and abreaction under hypnosis, I came up against two difficulties . . . which . . . led to an alternative both in my technique and in my view of the facts" (Freud, 1950, pp. 256, 262, 270). The two "difficulties" were (1) not everyone can be hypnotized, and (2) symptoms removed by hypnosis can and sometimes do recur. He chose to develop a system of therapy (psychoanalysis) without these limitations which he hoped would be of more universal use.

CASE 4. "One of my most acquiescent patients, with whom hypnotism had enabled me to bring about the most marvelous results, and whom I was engaged in relieving her suffering by tracing back her attacks of pain to their origins, as she woke up on one occasion, threw her arms round my neck . . . From that time onwards there was a tacit understanding between us that the hypnotic treatment should be discontinued . . . I felt that I had now grasped the nature of the mysterious element that was at work behind hypnotism. In order to exclude it, or at all events to isolate it, it was necessary to abandon hypnotism" (Freud, 1953, p. 27).

These are, of course, the words of Freud, describing an unexpected complication he encountered in hypnotizing a female patient and which led him to believe that hypnosis symbolized sexual seduction. Those familiar with the history of hypnosis will recall that Freud's mentor, Josef Breuer, had a similar experience with his patient "Anna O." Neither Breuer nor Freud estimated risk potential for hypnosis complications but both described their experiences with unexpected, unwanted side effects. When Freud first published his account of hypnosis complications in 1905, other authorities were urging caution in its use. Oppenheim (1904), in his widely used textbook on diseases of the nervous system, advised that hypnosis may "act injuriously and call into existence a severe hysteria so that great caution is necessary in its use. For this reason, if for no other, only experienced persons should use it, and then only for therapeutic purposes, and after other methods have failed" (p. 747). Bramwell (1903), a close friend and colleague of James Braid (1795-1860) who coined the term hypnotism, felt that "hypnotism, through ignorance or malice . . . might be so misused as to do harm" except "by medical men acquainted with the subject which then was devoid of danger" (p. 427).

Frederick Bjornstrom (1833-1889) was professor of psychiatry at Stockholm Hospital and cautioned in 1887 that "by suggestion of such a kind that with or without the operator's intention they can cause injurious and even fatal effects" (Bjornstrom, 1970, p. 411). He felt hypnosis "resembles a pathological rather than a physiological state" more "morbid and abnormal than as a healthy and normal condition" and, "in many cases indeed it is more like a mental disease than physical health and sound mind" (p. 410). Bjornstrom reported cases of hypnosis complications from 1784 to 1884, in England, Scotland and France, and how children in India in the early 1800s were hypnotized by kidnappers.

In recent years there has been an increase in the number of reports published in clinical and research texts and articles of unwanted, unexpected side effects coincident with the use of hypnosis. There range from severe life threatening emergencies, personality changes or psychotic-like decompensation, to mild, transient unpleasant throughts or feelings that fade within minutes. Frequently reported side effects are headaches, anxiety, irritability, fatigue, depression, unexplained weeping, dizziness, disturbed sleep or dreams, fear or panic attacks, lowered stress threshold, poor coping skills, depersonalization, derealization, disorientation, obsessive ruminations, delusions, psychomotor retardation, impaired or distorted memory, attention, comprehension, or concentration difficulties, sexual or antisocial acting out, and symptom exaggeration or substitution (Coe & Ryken, 1979; Conn, 1972; Faw, Sellers & Wilcox, 1968; Fromm & Shor, 1972; Hilgard, J., 1974; Kline, 1976; Levitt & Hershman, 1962; Meldman, 1960; Orne, 1965; Rosen, 1960, 1962; Sakata, 1968; West & Deckert, 1965; Williams, 1953).

Stage entertainers are involved in more reports of complications and with greater severity than clinical or research hypnotists (Echterling & Emmerling, 1982; Erickson, 1962; Gravitz, Mallet, Munyon & Gerton, 1982; Kleinhauz & Beran, 1981, 1984; Kleinhauz, Dreyfuss, Beran, Goldberg & Azikri, 1979). Hypnosis as used in experimental research appears to cause somewhat fewer complications than in clinical applications (Hilgard, J., 1974; Hilgard, Hilgard & Newman, 1961; Sakata, 1968; Williams, 1953).

CASE 5 was a 16-year-old girl who experienced 150 spontaneous trances in 30 days following being hypnotized on stage by an entertainer. They ceased only when she was rehypnotized by the same stage entertainer (Marcuse, 1964).

CASE 6 involved two 15-year-old girls hypnotized during a hypnosis "show" at their school. After the performance one of the girls "collapsed." She was rehypnotized by the same hypnotist and she was able to return to her home without further difficulty. The next day, however, both girls experienced spontaneous trance for periods up to 45 minutes. Both were rehypnotized and the spontaneous trances did not recur (Marcuse, 1964).

CASE 7 was a 24-year-old woman with no prior referral or treatment for mental problems and in good physical health. She was hypnotized on stage, suggested she was "rigid as a steel rod" and suspended between two chairs by her head and feet and the hypnotist then stood on her stomach. That night she had severe neck and back pain. She was hospitalized seven days and required orthopedic and psychiatric treatment for six months (H. Clagett Harding, 1977 and Milton Erickson, 1962 cited similar cases).

CASE 8 was a young man who was hypnotized on stage and told that an onion he was asked to bite into and eat was as delicious as an apple and he would really enjoy it. Inadequately dehypnotized he continued to enjoy the taste of raw onions. This compulsion persisted for five years until he was successfully dehypnotized by a psychiatrist experienced in hypnosis. While this case presented no medical emergency, the young man experienced social alienation from female peers as the result of this continuing posthypnotic suggestion (Prem, 1985 Toronto ISH Conference).

This case involved posthypnotic suggestion altering taste and smell and inadequate dehypnosis. The following case describes hypnotically induced tactile hallucination, altering the sense of touch and the termperature of an object:

Case 9 was a 30-year-old woman hypnotized on stage and told the hypnotist would hand her "a red hot poker." What he did was place a broom handle in her hand. The second it touched her she threw it to the floor, opened her eyes, cursed at the hypnotist and slapped him in the face. The audience applauded, thinking this was part of the show but the hypnotist was relieved when she stormed off stage and returned to her seat (Author).

Such immediate awakening is not unusual when subjects are suddenly or forcefully confronted with what they perceive as "a noxious stimulus" (Meares, 1960). Others accept or adjust to it, and still others do so with severe after effects. Another common source of hypnosis complications arises when individuals with no training experiment or "tinker" with hypnotic methods:

CASE 10 was an 18-year-old female student in a science project at school, experimenting with hypnotically enhanced responses to the blind Duke ESP card sort. A fellow student hypnotized her "after reading a book on it" and the experiment was completed. That night she experienced increasing anxiety culminating in a strong urge "to destroy the TV set." She went to bed but could not sleep, then suddenly jumped out of bed, "dancing, twisting and calling out, 'I'm an indentured servant.'" She was hospitalized four days and required psychiatric treatment for two months. She had no prior history of mental disorder (H. Clagett Harding, 1977).

Experimenting with hypnosis without training or supervision can be very dangerous, as this case clearly demonstrates. The same risk is involved with autohypnosis audiocassettes which many hypnotists provide their subjects. The danger is when and how the subject chooses to use them. In most cases, subjects use the tapes when they are alone, and, if they experience complications or side effects, there is no one with them or available to them to help them or arrange for prompt professional intervention. There is also a great temptation of medical and graduate students to test their skill on friends, individually or at parties. Gravitz, et al. (1982) cautioned that hypnotists should "take steps to ensure that those who learn hypnosis techniques . . . do not misuse what they might have acquired by practicing on themselves or others" (p. 307).

EVALUATING RISK

"In evaluating dangers, " Orne wrote in 1965, "it is necessary to take into account the context in which hypnosis is employed." There are risk factors common to all hypnosis settings such as subject personality characteristics (Chapter 2 of this book), those involving personal and professional behaviors of the hypnotist (Chapter 3) and those in the physical environment or treatment setting (Chapter 4). There are risk factors which are unique to the setting, hypnosis as stage entertainment, in clinical treatment and in experimental research.

Major risk factors in **stage hypnosis** are the lack of prescreening and very weak informed consent. If subjects are informed or led to believe hypnosis is "always safe" or "just fun" they can be lulled into a false sense of security, increasing their vulnerability and the risk of uncovering mentally "unfinished business" or personal problems, concerns or sensitivities clumsily stumbled upon and exposed publicly without professional help or support. What could be a therapeutic breakthrough in the confidential setting of hypnotherapy can be antitherapeutic trauma in stage hypnosis. Even without such dramatic after effects, the misconception that hypnosis is all-powerful can leave subjects who fail suggested tasks feeling inadequate and inferior. If it is believed hypnosis and the hypnotist never fail and there is failure the subject is likely to feel at fault.

As the cases in this book demonstrate, incomplete or ineffective dehypnosis greatly increase risk of complications. The larger the audience and number of subjects on stage, the less time and opportunity to observe for unexpected, unwanted effects. Usually no follow up is provided to discover and treat any lasting untoward effects. The entertainer moves on to the next engagement. The audience which sat anonymously in the semidarkness of the theater or night club, seeing and hearing several hypnotic inductions and posthypnotic suggestions, go home unaware of the effect these experiences may have on them.

The **clinical setting** in some ways is the opposite of the theater. It is private, not public, individual and not a group or shared experience, more personal and serious than social and frivolous. Each subject brings to this setting a distinctive perceptual or mind set, a complex mosaic of all the person is and has been, the obvious and the hidden, the conscious and the unconscious. The clinical hypnotist's ability to minimize risk and promptly and appropriately intervene as complications develop is a function of skill and judgment based on the quantity and quality of training and supervised experience.

Group hypnosis, in therapy or for habit control, presents some risk factors of stage hypnosis but is diminished by the skill, judgment and experience of the hypnotist. Some subjects in a group setting complain of difficulty achieving trance because the shared, group induction is so generalized to fit all it lacks individualization. Some complain about distractions such as coughing, clearing the throat, even shifting position. But the major risk factor is the hypnotist's difficulty in observing and monitoring everyone at one time — a lack or loss of control. A subject experiencing difficulty in group hypnosis may not be detected as quickly as in an individual session.

Autohypnosis with or without audiocassette tapes is without supervision, done individually when subjects are alone. Subjects should be carefully instructed in its use and what to do if for any reason problems or unusual after effects are experienced.

Lay hypnotists lack the years of education, training, clinical internships, supervised experience, licensing requirements and professional standards of care of psychiatry, clinical psychology, medicine or dentistry, where all these prerequisites are met before specific hypnosis training and supervised hypnosis experience. By their previous education, training and supervision in their respective professions, physicians, psychologists and dentists have far more likelihood of exercising seasoned judgment than those with lesser degrees and training.

Hypnotists in **experimental research** function in a more carefully controlled "laboratory setting" than clinical hypnotists. The experiment itself also offers structure and control. There is usually less time for the closeness and spontaneity of clinical practice. Risk factors unique to the research setting are the experimenter's lack of clinical skills, especially psychodiagnostics and psychopathology, weaknesses in the experimental design which allow for and may inadvertently facilitate complications, experimenter effects, unintentional cues which subtly or directly shape subject responses or intrude upon or become involved in the subject's mental state or personality dynamics beyond the scope or intent of the experiment.

INCIDENCE OF COMPLICATONS

Levitt and Hershman (1962) got a 34.6% return or 866 of 2500 questionnaires sent to members of professional hypnosis associations and others known to use hypnosis, asking them to report "unusual reactions" to hypnosis. Only 301 of the responses were sufficiently detailed to analyze or 34.8% of the 866 returned questionnaires. Results were as follows:

Unusual Reactions Rank Ordered by Percentile Incidence

(n = 301)

Unusual Reaction	n	%ile
Emotional upset (anxiety, panic or depression)	29	9.63
Headache, vomiting, fainting	15	4.98
Crying and hysteria	9	2.99
Loss of rapport during hypnosis	7	2.33
Overt psychosis after hypnosis	5	1.66
Difficulties involving sex	5	1.66
Excessive dependency on hypnosis	3	1.00
Difficulty resulting from inadvertent suggestions	3	1.00

A total of 76 "unusual reactions" were reported which yields a casualty rate of 25.2%

In 1974, Josephine Hilgard studied hypnosis after effects in 120 student volunteers screened by both Harvard (1962) and Stanford (1962) hypnotic susceptibility scales. Many after effects were observed during hypnosis. Headaches, confusion and anxiety sometimes faded when the hypnosis induction or session ended but some persisted. Depth of trance correlated weakly with intensity of after affects. Perceived body and self image were sometimes distorted in these otherwise healthy "normal" college studets, typified by the female subject who reported: "My body got bigger and softer, at the same time the actual me was becoming smaller" (p. 285, 286). The following table summarizes results of this study:

Hypnosis After Effects

(n = 120)

Symptomatology	Short-term (5-60 mins)	Long-term (61 mins +)	TOTAL
Headaches	1	3	4
Dizziness or nausea		1	1

(table *continued*)

Symptomatology	Short-term (5-60 mins)	Long-term (61 mins +)	TOTAL
Stiffness of arm(s)		1	1
Stiff neck	1		1
Drowsiness or sleep	8	7	15
Cognitive distortion and confusion	8	2	10
Anxiety	1		1
Dreams (night)		4	4
TOTAL AFTER EFFECTS	19	18	37
PERCENT OF TOTAL SAMPLE	16	15	31

Drowsiness and confusion were the most frequent short-term after effects occurring in 16 of 19 cases. Nine of these faded in 5-15 minutes, four persisted 15-30 minutes and three 30-60 minutes. Hilgard suggested these symptoms may be "a continuation of hypnosis rather than a sequel to it" (p. 288). Long-term after effects "were essentially continuations of . . . short-term after effects" in 9 of 18 cases but six of these persisted up to three hours and three others lasted much longer.

Dr. Prem, a psychiatrist in Scotland, reported at the 10th International Congress on Hypnosis and Psychosomatic Medicine in Toronto, Canada in 1985 sixteen casualties referred to him after a stage entertainer's performance at a local theater. While there were 2000 persons in the audience it was not possible to determine how many of these suffered side effects. Some may not have associated symptoms with hypnosis. Others may have consulted family physicians or mental health workers unknown to Dr. Prem. The sixteen persons complained of headache, dizziness or nausea which persisted up to a year after the hypnosis experience. Three of the sixteen decompensated into atypical psychosis, going into spontaneous trance.

Echterling and Emmerling (1982) polled 105 college students who attended a hypnosis show on campus. Of these, 32 or 33% had what they perceived to be a "negative experience." Echterling and Emmerling observed that "the majority of the audience will enjoy watching" and about "half of the individuals who have a trance experience during a stage hypnosis program will generally enjoy it." As for unwanted, unpleasant after effects, "about one fifth of its trance subjects will have both a negative experience during the trance and some negative after effects" and "about one in fifty of the other audience members also will experience negative consequences as a result of watching the program." This points up a danger unique to stage hypnosis: the large number of persons in the semidarkness of the audience unaware they may be partially in trance. "Several of the audience responded to virtually every hypnotic command regardless of whether it was directed towards the audience or a person on the stage," Echterling and Emmerling reported, concluding: "Stage hypnosis, in our opinion, poses dangers that are unacceptable and outweigh its potential entertainment value . . ." (pp. 10-13).

From the foregoing, the percentile incidence of hypnosis complications was:

INVESTIGATOR	POPULATION	n	%ile
Levitt & Hershman	Clinical	301	25.2
Josephine Hilgard	Experimental	120	31.0
Echterling & Emmerling	Entertainment	105	33.0

These three independent studies represent a sampling of hypnosis subjects in the three major settings: clinical, experimental and stage entertainment. It is unwise to generalize from one study of each of these populations but they involved large samples and were in this author's opinion, well designed and conducted. The collective data suggest a one in four chance of complications in clinical and research settings and one in three in the stage entertainment setting. Anecdotal case studies describe more severe complications from stage hypnosis.

Meares (1960) held that more complications occur in the clinical than the research setting. Barber (1961) concurred, suggesting that "the lab is not the real world." Orne (1965) pointed out the time constraints and the impersonal businesslike or mechanistic nature of research. More personal interaction facilitates deeper emotional involvement and "emotional aspects of the material covered make therapy a highly affect-laden environment in which reactions to hypnosis may be strong" (Gravitz et al., 1982, p. 305). On the other hand, clinicians are more apt to observe and intervene as complications develop (or could and should do so). They share more time with subjects and have a more complete history — a better database of personal dynamics. We need more large scale research in all three hypnosis settings before we can definitively state the risk potential for each. It is the author's bias that the need for training in preventive practices and risk management should be an even higher priority.

Risking oversimplification, researchers are not always aware of clinical realities and clinicians tend to lack a researcher's methodical preci-
sion. We are in desperate need of both. We do not yet have a clear
definition of what complications are and no standards of classification or
reporting. There are too few studies of large numbers of subjects and
those we have lack detail as to type of symptom, comparative severity
and exact duration in time. Despite these limitations, the following esti-
mates of complications risk are offered:

Population	Mild Intensity	Moderate Intensity	Severe Intensity	Total Casualties
Subjects in research	15%	5%	2%	22%
Clinical clients/patients	15%	5%	2%	22%
Stage hypnosis subjects	20%	10%	5%	35%

"Any discomfort" includes any unexpected, unpleasant or unwanted
after effects. One of every twenty subjects hypnotized on stage could
have severe after effects. One in fifty clinical or research subjects would
be likely to experience severe complications. If preventive practices were
routinely used and a system of risk management in place the author esti-
mates these percentiles would be reduced by half.

HYPNOSIS "ACCIDENTS"?

The problem situations described so far have been unpleasant and
unwanted by both the hypnotist and the subject. They were also unexpected, occurring suddenly and without warning. In these respects they
are analogous to auto accidents. Like auto accidents they are difficult to
analyze in such a way as to provide helpful preventive practices for
drivers (for this analogy, hypnotists). When a car is "totalled" (the subject hospitalized) there can be no doubt an accident (hypnosis complication) has occurred. A small dent, chip in the windshield or "fender
bender," where no one is hurt and the damage less than insurance de-
ductible, are often repaired without police or insurance claims. These
are minor accidents (mild complications). Injuries can be mild, moderate or severe, acute (short term) or chronic (long lasting) and so also in
hypnosis "accidents."

The most expensive car (best training) and the most experienced driver (hypnotist) can be involved in an accident because of the attitude and behaviors of others (hypnotic subjects). Not all drivers know what to do in an emergency. Driving too fast or too slow, unfamiliar with the road or conditions make accidents more likely. Carelessness, negligence, impaired function due to alcohol or drugs, can cause accidents with cars, with hypnosis, in medicine, psychology, or dentistry, or any other profession or trade.

Most professional hypnosis courses take less time than the average high school driver education course and involve less supervised experience. Worse still, in most states, provinces and nations, no license is required. You can be illiterate or a convicted felon. You don't have to even read a book on hypnosis to practice it. You can learn by doing without any education, training or experience whatever. It should come as no surprise that hypnosis complications are likely to occur more frequently than auto accidents.

DEFINITIONS

There is a critical need to define terms so that a more valid and reliable database can be accumulated to further clarify the nature, frequency, and severity of hypnosis complications. Then, effective preventive practices can be developed and implemented as standard practice. The following definitions are offered:

Hypnosis

Basic to any study of hypnosis complications is an understanding and definition of hypnosis itself. Hypnosis is not yet precisely defined to the satisfaction or agreement of all who use it. There is no clear consensus as to what in fact it is or is not. Some define it in very broad terms including daydreaming, reverie, total absorption in a hobby or pastime, political, religious and sexual experiencing, and similar "turning on" as hypnotic processes. Others define hypnosis as an altered state of consciousness, a unique subjective state, dissociation or neodissociation, a complex interaction of many variables, right hemisphere brain function, cortical inhibition, subcortical function (brain stem or limbic system), primitive instinct or archaic involvement, regression to childlike trust or dependency or imaginativeness, or simply increased suggestibility. In

1903, Bramwell offered a practical, working definition as timely today as then, that when a suggestible person becomes more suggestible, that difference is hypnosis.

We understand and use hypnosis as we understand and use electricity, gravity and intelligence — phenomena which exist but are also not precisely defined. The definition preferred by hypnotists usually reflects a theoretical orientation, earlier training emphasis and expectations of subjects. The hypnotist projects these when explaining hypnosis to subjects, obtaining informed consent, and conducting the induction. Movies, TV, books and magazines can in effect program subject expectations and trance behaviors: "If specific behavior becomes widely publicized as characteristic of hypnosis, subject and hypnotist alike may view it as typical. It will then tend to occur when hypnosis is induced" (Orne, 1962b, p. 681).

Complications

Complications are unexpected thoughts, feelings or behaviors during or after the hypnosis which are inconsistent with agreed goals or the subject's expectations, interfere with the hypnotic process and threaten optimal mental functioning and the person's mental state. There is no prior history of treatment for them or similar mental or physical problems. Usually perceived as unpleasant, they are nontherapeutic (would not be part of a clinical treatment plan or research protocol) or antitherapeutic (they leave significant treatable psychopathology clinically equivalent to post-traumatic stress disorder). Intensity varies (mild, moderate, severe), from unpleasant, annoying, irritating or uncomfortable symptoms which fade with or without the hypnotist's intervention (mild intensity) to those which persist and cause agitation or disruption to the subject's lifestyle or life situation (moderate intensity) to totally disabling or life threatening emergencies (severe intensity). Duration is acute (short term) lasting minutes, hours, days or weeks, or chronic (long term) lasting months to years.

Hypnosis complications can be relieved by **rehypnosis** and this is a distinguishing feature and a diagnostic sign. The following summarizes the distinguishing characteristics of hypnosis complications:

Complications are

    unexpected and unwanted
    thoughts, feelings, behaviors

perceived as unpleasant
  or of crisis proportion
nontherapeutic or antitherapeutic
mild, moderate or severe
  (mild: fade without help;
  moderate: disrupts life;
  severe: totally disabling)
acute or chronic
  (acute: last minutes to weeks)
  chronic: last more than a year)
onset sudden or delayed

Preventive Practices

These are specific interventions and techniques to prevent complications or to promptly intervene when they arise.

Risk Management

This is an organized, systematic process of standard conditions and procedures to reduce risk of complications (e.g., standard forms for history and informed consent; checklist format for screening, observation and induction; written progress notes)

In addition to the lack of definitions, there are other major obstacles to implementing preventive practices and risk management. Hypnosis is not "owned" exclusively by any of the mental health professions. It is an orphan shared by many, without parents, with no one "responsible" or "in charge" to enforce standards of competence and practice. There is no legal control restricting or regulating its use in most states, provinces and nations (Marcuse, 1964). Sweden has had legal restrictions on the practice of hypnosis since 1906. Ontario, Canada has had regulations for more than twenty years.

SUMMARY

Lacking a universally accepted definition, without legal constraints, complications are likely to occur with such severity they persist for years. Frankel (1976) observed that "any method of therapy, including therapeutic hypnosis, is likely to succeed only when it is the right treatment, at the right time, and can mesh with the psychodynamic forces at work in the clinical picture" (p. 133). To this recommendation Weitzenhoffer (1957) added his opinion that complications and adverse effects are not caused by hypnosis but rather by the hypnotist, depending on "the competence and integrity of the practitioner."

In 1962 the Group for the Advancement of Psychiatry issued a position paper entitled *Medical Uses of Hypnosis* which stated:

> Whoever makes use of hypnotic techniques . . . should have sufficient knowledge of psychiatry, and particularly psychodynamics, to avoid its use in clinical situations where it is contraindicated or even dangerous. Although similar dangers attend the improper or inept use of all other aspects of the doctor-patient relationship, the nature of hypnosis renders its inappropriate use particularly hazardous . . . more than a superficial knowledge of the dynamics of human motivation is essential (p. 705).

Untrained, inexperienced or inept practitioners are more likely to cause complications and side effects because they are less aware of risk factors, not as trained to observe, interview, and take a careful history, and are more apt to miss cues, misinterpret, or misdiagnose serious psychopathology. As Meares pointed out (1961): "The induction of hypnosis by an unskilled person can represent a real danger to the subject." What the hypnotist says and does, how it is said and done, can impact the subject more in trance than in the unhypnotized awake state: "Things seem to have a strange symbolic meaning" (Meldman, 1960).

The hypnotic subject may not be the "normal" well adjusted person assumed by the hypnotist as will be demonstrated in the next chapter. Subjects bring to hypnosis all the many variables which make up personality, remembered and unremembered significant events in their lives past and present, and any of these can erupt in a crisis situation during or after hypnosis. Underlying "unfinished business," characterological or psychotic disorders can rise to the surface when already weak ego defenses are further relaxed by hypnosis. Themes of aggression or of submission can emerge, differing dramatically from how the subject might want to react, raising questions of human rights and informed consent further discussed in Chapter 5.

The environment (the treatment or experimental setting) and the materials, equipment and furnishings there, can become risk factors. Subjects not debriefed to white sound can go into trance when hearing similar sounds (MacHovec, 1986). Researchers may have difficulty replicating experiments because of unreported differences in room size, acoustics, furniture and its placement, time and timing, even the time and day, temperature and humidity. Tight clothing, body position and physical touching can lead to perceptual distortion.

CONCLUSIONS

Hypnosis is intrusive to the subject's mental state and "is not entirely innocuous" (Meldman, 1960). It directly and indirectly involves the subject's thought and perceptual processes, affect and memory. The usual pre-induction relaxation suggestion encourages a relaxation of critical judgment, rendering the subject more vulnerable to "an intense interpersonal encounter" (Kleinhauz & Beran, 1984).

These data suggest that hypnosis complications are underestimated, unrecognized and underreported in both frequency and severity. The paucity of published reports of casualties, the lack of definition, standardization and control contribute to this situation. The general public lack awareness of the risk factors and frequent after effects and are therefore less likely to associate symptoms with hypnosis. This in turn can delay treatment. Some severe complications have gone untreated for years.

Hypnotists need to increase their awareness of risk factors in subjects, in themselves and in treatment and research settings. Added care should be taken to make hypnosis as safe and risk-free as possible, establishing standard laboratory conditions. Most hypnotists have been taught that hypnosis is safe and if they have not observed serious complications, may resist adopting preventive practices. Thomas Huxley wrote in 1880 that "irrationally held truths may be more harmful than reasoned errors." If hypnosis is as safe as we have been led to believe in courses, books and articles in the past, this book is a reasoned error. If it is not, refusing to consider and use preventive practices is blind faith in an irrational truth.

Mesmer himself, looking back over the difficulties he faced and the rejection he experienced, showed keen insight into why differing viewpoints (such as dangers of hypnosis) are resisted: "It is not easy to renounce accepted ideas, the principles of one's education, the efforts of one's youth, the reputation one has made growing old. These sources of resistance seem to me the true enemies of animal magnetism" (Shor & Orne, 1965). It is hoped rational truth will emerge in the search for the nature of hypnosis: what it is, but also what it should be. May it be as Socrates suggested 2500 years ago:

There is only one good — knowledge
And one evil — ignorance.

CHAPTER 2

SUBJECT RISK FACTORS

Things are seldom what they seem;
Skim milk masquerades as cream.

— HMS Pinafore, ACT I
(Gilbert & Sullivan)

THESE two lines from Gilbert and Sullivan's comic opera *HMS Pinafore* epitomize the risk factors for possible complications a subject can bring to the hypnosis situation. These factors are "so varied and subtle that a great variety of behaviors can be theoretically expected from the process" (Williams, 1953, page 3). This "great variety of behaviors" involves the subject's personality dynamics, previous life experiences (real or imagined), current mental state and life situation. Unresolved conflict or "unfinished business" involving any of these major variables can emerge during or after hypnosis as part of what Meares (1960) termed "traumatic insight." If unexpected and unwanted, they are complications and they occur in all hypnosis settings: clinical practice (habit control, psychotherapy), experimental research, and entertainment or stage hypnosis.

This chapter will consider subject risk factors likely to interfere with hypnotic induction or precipitate complications during or after hypnosis. These factors are: personality traits; attitude and expectations; inappropriate indirect methods; paradox; attention, comprehension or intellectual deficits; underlying characterological problems; prepsychotic states; secondary gain; resistance; symptom removal, substitution or exaggeration.

The person who agrees to be hypnotized brings to the hypnosis situation the totality of personality resulting from an infinity of life expe-

21

riences, happy and sad, real and imagined, "a complex matrix of interacting variables" (Barber, 1961, p.119). Unless the hypnotist obtains a good history and carefully observes the subject before, during and after hypnotic induction, there is increased risk that repressed material, the "unfinished business of the mind" will unexpectedly emerge. Something said or done can evoke conscious or unconscious reactions in defense of the subject's self concept or value system. To proceed without an awareness and knowledge of subject risk factors is much like attempting to play cards with an incomplete deck. Some subjects respond as if they held more cards than the hypnotist. Worse still, some few respond as if they were playing with a marked deck!

SUBJECT PERSONALITY TRAITS

Meares (1960, 1961) described several personality traits which can interfere with the hypnotic process and contribute to complications:

Overdependency, where the person seeks through hypnosis to "satisfy inner needs." Meares cautioned that "the over dependent patient may be made still more dependent." Overly dependent subjects do not benefit sufficiently from hypnosis to achieve autonomy or self-reliance. They usually phone for "a recharge" of hypnosis or seek treatment for additional symptoms or problems.

CASE 11 was a 39-year-old man who was treated with hypnosis for habit control, to stop smoking. Progress was slow and while he experienced relaxation using autohypnosis, his rate of smoking decreased very slowly. At about the time he could avoid smoking completely he asked that hypnosis continue for weight loss. This, too, was a slow process and when it was near completion he asked hypnosis continue for drinking. He then asked that posthypnotic suggestion include improving his memory to help him in his night school courses. The hypnotist realized his dependency at about the time he inquired if he could "try hypnosis for age regression to see who I was in a previous life" (Author).

Masochism, seeking to satisfy a wish or need to be overpowered. There are women who may seem quite assertive and independent but who satisfy an unrealized need to be dominated. There are men who are successful managers, active, decisive and self-reliant, who use hypnosis to "lower" themselves and seek "to feel humble . . . without driving ambition . . . out of the rat race." These men and women verbalize a wish for someone to "take over for a while" so they do not have to be as vigilant and "in charge."

CASE 12 was a 36-year-old woman who was married three times. All three husbands were alcoholics and wife beaters. She was referred to hypnosis treatment to reduce anxiety and improve coping skills. Hypnosis was successful. She was able to use autohypnosis to reduce tension. She continued in her marriage and continued to be physically abused by her husband. In effect, hypnosis helped her adjust to a situation in which she chose to remain, even more submissive than before (Author).

CASE 13 was a 27-year-old woman who sought hypnosis to help her "be more assertive." Her husband verbally and physically abused her. She responded well to hypnosis, felt more relaxed "being herself" and gradually began to feel more comfortable "speaking her mind." As she became more assertive, her husband became more violent toward her. The hypnotist hoped she would so realize her individuality and with her newly found assertiveness leave her husband. Instead she required hospitalization from increasingly serious injuries, including a broken jaw and a fractured skull (Author).

Aggression, where the individual feels the need to "get even" or "pay back" or "show" others. For such men and women the hypnotist is symbolized as the targeted person(s), perhaps because s/he looks or sounds like one's parent, former teacher, boss or other authority figure. Some women will use the hypnotic session to "get even" with males, through resistance or venting hostility during the hypnotic induction. Some men, thwarted by their job or family ties from venting their true feelings, will do likewise. If the hypnotist and subject are of the same sex the interaction might symbolize peer or sibling conflict.

CASE 14 was a woman treated successfully with hypnosis for "severe generalized pruritus" but as her physical discomfort decreased she became more and more agitated. She became "almost overwhelmed with sexual desire" for her lover but at the same time she paradoxically experienced "an almost homicidal rage against him." Her "itch" symbolized her ambivalence toward her lover and she probably would have benefitted more by individual psychotherapy for the strong emotions beneath her physical symptoms (Rosen, 1960, p. 241).

Prepsychotics may use hypnosis to free themselves from the restrictions of family, society or even their mental state, and decompensate into psychosis during or after hypnosis. Meares felt there is a very real danger of precipitating a psychotic break by using hypnosis on a person who is at or near decompensation or who has an encapsulated psychosis.

CASE 15 was a man treated with hypnosis for phantom limb pain following neuronectomy. Hypnosis relieved his pain but "a short time later he was hospitalized psychiatrically because of an underlying schizo-affective psychosis." While it is highly unlikely hypnosis "caused" the psychosis its use coincided with and seems to have facilitated it (Rosen, 1960, p. 241).

CASE 16 was a man treated with hypnosis for the relief of pain following laminectomy. "His pain disappeared posthypnotically and the following week he committed suicide by jumping out of an upper story window." If his back pain had an organic cause, treatment only with hypnosis was not appropriate. If the pain was psychogenic, he should have received ongoing individual psychotherapy, possibly augmented with hypnosis. Rosen, who reported this case, observed that "pain frequently masks severe and even suicidal depression or serves as their equivalent" (Rosen, 1960, p. 241).

Erotic drives, realized or unrecognized, real or imagined, where women may become erotically aroused or emotionally involved in a loving relationship with a male hypnotist, or males with female hypnotists, or latent or overt homosexual attachment between those of the same sex. A frequent cause of unfounded charges of sexual advances or sexual contact between hypnotist and subject involve the subject's erotic drive. A careful study of his or her life situation usually provides evidence to substantiate this (e.g., recent divorce, social deprivation, other stressors).

CASE 17 was a male adolescent treated for "wry neck" (torticollis). Hypnosis had little effect. He was age regressed and relived "a sexually traumatic episode" after which the torticollis disappeared. But he experienced a sudden, dramatic personality change and "became an exhibitionistic practicing homosexual." It may be that he would have "come out of the closet" anyway, independent of any hypnotic treatment. But his "break" from his previous lifestyle occurred suddenly and coincided with the use of hypnosis. It can also be argued that it would have been more therapeutic to have provided him with ongoing psychotherapy over a longer period of time to help him explore his sexual identity and preference, allowing his decision to evolve more calmly and naturally (Rosen, 1960, p. 142).

Meares observed that some clients dress or behave in an obviously seductive manner, a certain sign to the wary hypnotist of potential danger. A strong underlying erotic drive can be disinhibited by hypnosis, as Meares pointed out, dramatically demonstrated in the classic case of Josef Breuer, Freud's mentor, when a female patient suddenly and unexpectedly threw her arms around him. For Meares as well as Freudian therapists, "latent homosexuality" can through hypnosis become "manifest" in "homosexual panic." It is also true that some, fortunately few, hypnotists intentionally or unintentionally respond erotically to certain of their clientele they find sexually attracted.

The DSM-III *(Diagnostic and Statistical Manual of the American Psychiatric Association,* 1982) provides a more comprehensive description of personality disorders which may obstruct and interfere with the hypnotic process. For convenience here and at the risk of oversimplification, personality disorder is defined as a long standing (usually lifelong) consistent, repetitive, distinctive pattern of behavior. When these patterns are less pronounced, less severe, they are considered to be personality traits, not personality disorders. DSM-III personality traits/disorders and their likely effect on hypnotizability:

Paranoid (301.00) where the individual's suspiciousness, distrust, hypersensitivity or emotional distracing can obstruct the hypnotic process;

Histrionic (301.50), formerly known as and still referred to by some clinicians as hysterical personality, where the subject is so dramatically over-reactive, intense, though really superficial, and so self-indulgent, demanding or manipulative as to grossly exaggerate hypnotic suggestion or use hypnosis to further reinforce existing or exaggerated behaviors;

Narcissistic (301.81) where grandiose pre-occupation with one's self and an incessant search for self insulates subjects from "normal" thoughts and feelings and makes them inaccessible to hypnotic suggestion or they screen the suggestion retaining only content that bolsters their already inflated self esteem;

Antisocial (301.70) and also **borderline personality (301.83)** where a selfish disregard for others and the consequences of their acts, can render them so resistant or defiant as to prevent full hypnotic effect;

Avoidant (301.82), hypersensitive to rejection, with a resulting paradoxical defensiveness, therefore unable or unwilling to "let go" and fully cooperate. These subjects are usually socially withdrawn, have low self esteem, and crave affection and acceptance. Conversely, they may be very good hypnosis subjects, using it as a substitute for affection and acceptance;

Dependent (301.60) can be very good hypnotic subjects—too good. They easily lose themselves in the hypnosis process, passively relinquishing control—and responsibility—to the hypnotist, assuming less autonomy and less self-reliance than is suggested. There is a very real risk of a paradoxical reaction of further helplessness or total dependency on the hypnotist. It is possible to mistake this overdependency for a "strong transference";

Compulsive (301.40) where the subject is perfectionistic, overly attentive to detail, overly meticulous, straining to hear every word, trying too hard, overly upset by interfering thoughts and ideas, and by being so "bothered" will obstruct the hypnotic process. Meares (1960) predicted that compulsive persons who are hypnotizable are likely to have sudden panic reactions when they realized they were hypnotized. He felt some women might be erotically aroused to orgasm and some men overwhelmed by either regressive feeling toward a parent or authority figure or "homosexual panic."

Passive-aggressive (301.84), passive obstructionists who indirectly resist suggestion and sabotage the hypnotic process by being evasive, forgetful, unclear, inefficient, unable (unwilling!) to "put it all together." These subjects often arrive late, cancel or forget appointments, or terminate therapy prematurely;

Schizoid (301.20) where the subject is indifferent, aloof, withdrawn, "to himself" and bafflingly inaccessible, where all that is said and suggested is screened against one's own unique "self talk" and individualistic "inside language" and whatever doesn't match is ignored, not even heard. In this sense, these subjects are "not here";

Schizotypal (301.22) where an abnormally rich fantasy life such as excessive pre-occupation with ESP, altered states, superstition, religiosity, or bizarre thoughts and ideas, ideas of reference, social anxiety or withdrawal, or, in extreme cases, depersonalization, derealization, and vague, circumstantial or metaphorical speech, render them inaccessible to the desired or planned hypnotic suggestion. They are likely to use hypnosis as a vehicle for even more bizarre mental excursions with risk of a psychotic break;

Individuals can have any one or a combination of several of these traits. Those who have been smoking, drinking or eating to excess for years are apt to resist posthypnotic suggestion to control these habits, even though they pay for and agree to submit to treatment for them. In doing so, they are being both avoidant and passive-aggressive.

Multiple personality (300.14) is rarely encountered as compared to the above disorders and presents unique risks of complications. It is a dissociative disorder in the same DSM-III diagnostic category as amnesia, fugue and depersonalization but differs from them with the existence of two but usually more distinct personalities in one individual. Each identity takes over control of the individual at different times in certain trigger situations. The life history is often replete with severe stress, deprivation or traumatic incidents. Integrating that history and the multiple identities into an autonomous whole is the treatment goal and hypnosis has proven effective in facilitating this. Difficulties arise if rapport is lost with any of the personalities or if a malevolent one takes control. There is a very real danger of suicide if the dominant personality seeks to destroy the others, resulting in the death of the individual.

HYPNOSIS COMPLICATIONS AS MENTAL DISORDERS

Hypnosis complications, of and by themselves, can satisfy DMS-III diagnostic criteria as mental disorders. If subjects do not associate the symptoms with hypnosis, they can and often do consult a family physician or seek medical treatment for them. If the examining physician, psychiatrist or psychologist is not aware that the subject has been recently hypnotized, it is possible medical or psychotherapeutic treatment will be ineffective. Clinicians who use hypnosis should help educate non-hypnotist colleagues to the following DSM-III disorders which can be misdiagnosed hypnosis complications:

Adjustment disorders are maladaptive reactions which occur within three months of an identifiable stressor (hypnosis and/or thoughts, feelings, behaviors coinciding with it), resulting in impaired social or occupational functioning or overreacting to the stressor, not a recurrent pattern or part of another problem or condition, and which can be overcome when a new level of adaptation is achieved. There are several types:

Adjustment disorder with depressed mood (309.00) when depressed mood, tearfulness and hopelessness predominate;

Adjustment disorder with anxious mood (309.24) when anxiety, worry, or restlessness predominate;

Adjustment disorder with mixed emotional features (309.20) where anxiety, depression or other emotions are present;

Adjustment disorder with disturbance of conduct (309.30) where the subject violates age-appropriate societal norms or the rights of others;

Adjustment disorder with mixed disturbance of emotions and conduct (309.40), where there maladaptive behavior involves both emotion and conduct;

Adjustment disorder with work or academic inhibition (309.23) where the subject experiences problems at work or in school whose previous performance was adequate; usually there is a significant level of anxiety or depression (examples: writer's block; difficulty studying; test nerves);

Adjustment disorder with withdrawal (309.83) where social withdrawal without significant anxiety or depression is the predominant manifestation;

Adjustment disorder with atypical features (309.90) where symptoms do not fit the foregoing criteria.

Generalized anxiety disorder (300.02) where there is generalized, persistent anxiety for a month or more in a subject 18 years of age or older, and three or more of these:

> **motor behaviors** (tremors, shakiness, muscle pain, tiredness, eyelid flutter, furrowed brow, strained facies, fidgeting, low startle reflex threshold, etc.);
>
> **autonomic hyperactivity** (sweating, tachycardia, cold clammy hands, dry mouth, dizziness, light-headedness, paresthesias in hands or feet, upset stomach, hot or cold flashes, frequent urination, diarrhea, lump in the throat, flushing, pallor, rapid pulse and respiration, discomfort in pit of stomach);
>
> **apprehensive expectation** (anxiety, fear, worry, rumination, feeling of impending doom to self or others);
>
> **vigilance and scanning** (insomina, irritability, impatience, on edge, hyperattentive, easily distracted, difficulty concentrating).

Panic disorder (300.01) where the subject experiences three or more panic attacks within three weeks when not in a life-threatening situation or during marked physical exertion, and four or more of these:

- dyspnea
- palpitations
- sweating
- faintness
- paresthesias of hands and feet
- hot and cold flashes
- chest pain or discomfort
- choking or smothering feelings
- trembling or shaking
- fear of "going crazy"

- unsteady feelings
- fear of dying
- dizziness or vertigo
- fear of loss of control

Post-traumatic stress disorder symptoms occur when repressed material crashes through hypnotically-relaxed ego defenses. According to DSM-III criteria, the stressor (suddenly recalled trauma) is "recognizable" and "would evoke significant symptoms of distress in almost anyone." Reexperiencing the trauma is evidenced by one or more of these:

- recurrent, intrusive recall of the traumatic event;
- recurrent dreams of the event;
- sudden realization as if the event were recurring triggered by environmental or ideational cue.

There is also a "numbing or responsiveness to or reduced involvement with the external world" with one or more of these:

- marked disinterest in one or more significant activities;
- feeling detached or estranged from others;
- constricted affect.

And two or more of these which were not present before the trauma or hypnotic recall of the trauma:

- hyperalert or exaggerated startle response;
- sleep disturbance;
- survival guilt;
- impaired memory;
- difficulty concentrating;
- avoiding anything that would trigger trauma recall;
- intense symptoms in situations similar to traumatic event.

There are several varieties of post-traumatic stress disorder or PTSD as it is sometimes abbreviated:

Post-traumatic stress disorder, acute (308.30) where onset is within six months of the trauma or hypnotic recall of previous trauma and the symptoms fade in less than six months;

Post-traumatic stress disorder, chronic or delayed (309.81) where onset is more than six months after the trauma or hypnotically recalled trauma and/or the symptoms last more than six months;

Brief reactive psychosis (298.80): all diagnostic criteria for schizophrenia except for duration which is more than two weeks but less than six months.

Atypical psychosis (298.90), duration less than two weeks: perceptual distortion (hallucinations, delusions); confusion, disorientation or incoherence; loose associations, impoverished or markedly illogical thought processes; grossly disorganized or catatonic behavior. There can be delusions of body image distortion usually triggered by trance suggestions, or persistent auditory hallucinations (e.g., hypnotist's or other's voice heard inside the subject's head), grossly exaggerated recalled events, thoughts or feelings from elements of trance induction. Hypnosis can elicit a confusing array of clinical features.

Atypical bipolar disorder or "bipolar II disorder" (296.70): a less severe manic or depressive episode in a person who has previously experienced a major affective disorder of psychotic proportion.

Atypical depression (296.82): a distinct, sustained episode of classical deep depression in a person without recent psychosocial stressors or history of major depression and has had intermittent periods of "normal" mood for several months.

Clinicians who have had extensive experience using hypnosis are at one time or another likely to see all of the symptoms described above. The author has observed them all as sequelae to hypnosis, independent of prehypnotic presenting conditions or prior history, in subjects never before treated or referred for mental problems. It might be helpful to those clinicians and researchers who have not observed them to reflect over the subjects they have hypnotized or their treatment or experiment notes, to become more aware of their incidence and, hopefully, to develop and use preventive practices to avoid them.

ATTITUDE AND EXPECTATIONS

A **rigid moral code** or **strict value system** can predispose subjects to such side effects as headaches, insomina, restlessness or anxiety. The predisposing factor may be that hypnosis is perceived as a means by which their system might be weakened. Treated for depression, suggestions of elevated mood or pleasurable experiences may be misinterpreted as a step toward reckless abandon. Treated for habit control or a pain state, freedom from the habit or discomfort may suggest "freedom" from any and all constraints. If hypnosis "feels different," this can trigger a fear of "something new and different" somehow opposed to their values, morals or conscience.

CASE 18. A 42-year-old school teacher requested hypnosis for habit control. He wanted to stop smoking. He had no history of any mental problems and recent medical examination found no physical problems. After the first induction he complained of a headache. A second induction was then conducted including suggestions the head pain would diminish. It got worse. The pain was so great the subject terminated not only the session but the entire treatment plan. Pain persisted all that night requiring medication and a medical consultation the next day. Looking over the subject's history, the hypnotist noted that he was devoutly religious, a deacon in his church, with a very rigid value system. He "never had a headache in years" and the fact that he had none before or since hypnotic induction suggests that he found hypnosis or change, even of a bad habit, unacceptable. His value system was so rigid it would not allow for any change. Change to him literally was a real headache (Author).

CASE 19 was a female college student volunteer in a hypnosis experiment. She experienced a headache at the beginning of hypnotic induction which lasted for an hour and a half afterwards. She slept three hours and when she awakened the headache was gone. She described hypnosis as a stress situation for her "because I didn't want to get hypnotized and yet I was curious." A hypnotic dream was part of the experiment from the *Stanford Hypnotic Susceptibility Scale, Form C* (1962) and she described it thus: "I was in class. A lot of people around. A man was telling me about falling asleep. I was confused about what he was telling me; it didn't seem right." J. Hilgard commented that the dream "indicates some of her conflict about hypnosis . . . " (J. Hilgard, 1974, p. 281).

CASE 20 was a male college student in the same experiment described above. He also experienced a headache which began "toward the end of the hypnotic session" and lasted two hours "until he took a nap and got rid of it." He described his attitude toward hypnosis: "I really wanted it but it just didn't work." He said he felt compelled to be in control, not just in hypnosis but at any time. For this reason he would never drink to excess. "My mind keeps going," he said, "I don't blank my mind out when I'm trying to fall asleep. It was the same way in hypnosis. My body was relaxed, but not my mind." This compulsive attitude prevented him from experiencing the hypnotic dream item on the *Stanford Hypnotic Susceptibility Scale, Form C* (1962). "While being bothered that he could not dream he started to think of an earlier experiment that he had tried that did not work, a related conflict between aspiration and achievement" (J. Hilgard, 1974, p. 291).

CASE 21. The hypnotized subject in trance suffered "violent, uncontrollable trembling accompanied by nausea" when asked to mispell his name. He was then told he didn't have to comply if he didn't want to. The trembling increased! These unpleasant symptoms disappeared only when he actually misspelled his name. What the hypnotist judged to be a harmless task was not perceived as such by the subject who suffered obvious anxiety and discomfort. The supposedly simple task he intruded to a deeper level of the subject's consciousness. The misspelling became "a successful abreaction" for "an experimental neurosis unwittingly produced or a latent neurosis" (Williams, 1953, p. 4).

CASE 22. A dentist used hypnosis on a 36-year-old man to minimize pain and anxiety during routine cleaning and filling of two teeth. Asked if he "would like to go hunting" the patient replied that he would. The patient was not asked to suggest his own relaxing pastime. As the dentist described the visual imagery and sequence of the hunt, the patient became more anxious and uncomfortable. After the dental procedure was completed, the patient was asked about his obvious discomfort. "Yes, doc," he explained,"I was afraid I'd miss, get shot myself or shoot some other hunter by mistake." Had the dentist asked the patient for a calm and restful scene or pastime, this complication might have been avoided. The dentist's expectations (and perhaps his own projected need) of "a fun hunting trip" were not at all the same as the expectations of the patient (Author).

CASE 23 was a 14-year-old boy treated in the early 1960s for pain secondary to advanced periarteritis nodosa. Use of steroids caused Cushingoid symptoms, possibly severe vertebral demineralizaition, presumed to be the major cause of pain not relieved by high doses of narcotic analgesics. The boy was a dependent of a military family and he was referred to a military psychiatrist. He responded well, progressively, to hypnotic suggestion of relaxation and pain reduction. Medications were reduced with no intensification of pain. After two months' weekly hypnosis treatment, the psychiatrist was involuntarily sent for three weeks duty at another base. When he returned he found the patient not only experiencing pain, which returned within a week of the psychiatrist's departure, but he was unhypnotizable despite several attempts and a variety of induction techniques (Donald K. Jones).

Williams (1953) stated that hypnosis involves "significant personal participation in the process" and that the experience "will be fitted into a background of needs and objectives that is distinctive for each individual" (p. 7). The significance of what is said and done during hypnosis will vary with the individual in the ways described throughout this chapter, "even though a well-identified reason may be consciously given" by the subject. A person who seeks or submits to hypnosis for habit control, anxiety or pain relief, to facilitate psychotherapy, to volunteer in an experiment, or to "have fun" in a stage demonstration may be motivated by internal cues. The risk of serious after effects depends on what lies below the surface in the mind of the subject. If it is simply a natural curiosity or a mildly unpleasant childhood memory, all may be well. If involved in deeper crosscurrents of the mind there is high risk of a sudden, unexpected emergence of mental unfinished business.

CASE 24. Complications involving the subject's attitude and expectations can be both negative and positive. In this case a hypnotized subject who refused to come out of trance. The hypnotist forcibly opened the subject's eyes, even "gave him a vigorous shaking and took him by the nape of the neck." The subject replied: "There's something back of this. I don't know how but if you hypnotize me some more I think I'll remember—put me deeper." The hypnotist did so but the additional effort was not productive. Whatever was "back of this" resisted all attempts to expose it. This case illustrates how hypnosis can inadvertently involve deeper mental processes, figuratively like turning over a rock. For one person, there is little or nothing there. For another, there can be a lifetime of anguish uncovered with little or no preparation, relying on whatever strengths there are in that person and the hypnotist also involved in the uncovering (Williams, 1953, pp. 4, 5).

Faw et al (1968) offer further evidence that hypnosis involves deeper mental processes. They observed that in research studies involving "a normal college population" where no negative or detrimental effects were reported, experimenters observed "greater warmth and social interaction" in those hypnotized. They concluded "there may be a more complex process which was precipitated by hypnosis" (p. 35).

Whatever forces shape the individual personality affects how that person will react to external cues such as hypnotic suggestion. Among these are social learning, conditioning and reinforcement from parents, siblings and birth order, relatives, peers and significant others, race, sex, age group, culture, nationality and religion, marriage and child rearing, career, social status and rural-urban environment, intelligence, education and training, and traumas and illnesses. Even a physical symptom such as hand tremors can have a variety of causes such as an anxiety or panic attack, Parkinsonism, side effect from medications, or the effect of a crash diet.

Social psychology research provides valuable insights into attitude and motivation of volunteer subjects in experiments. Most are cooperative and supportive, believing that research and their part in it advances science (Orne, 1962a). Some are oppositional and negative, seeking indirectly to ruin the experiment. Argyris (1968) likened them to low-level employees of large organizations "dragging their feet" to fight back at management. Masling (1966) termed this indirect defiance "the screw you effect."

CASE 25. A woman referred to autohypnosis for anxiety reduction paid for and received six hours of hypnotherapy with little success, then confided she drank 30 cups of coffee a day. Asked why she hadn't reported this before, she smiled naughtily and said: "Because you didn't ask me." Her "cat and mouse game" had more importance to her than agreed treatment goals and winning that game or "one upping" the therapist was of a higher priority than relieving her chronic anxiety (Author).

Ambivalence. Shevrin (1972) reported that some subjects experience coexistent alternating positive and negative thoughts or feelings toward hypnosis or the experimental research situation. J. Hilgard (1974) described this like-dislike dilemma as a "conscious enjoyment of the experience but a distaste or fear of the experience at another level" (p. 295). Williams (1953) observed that "in some cases what were meant to be innocuous suggestions of well-being and happiness were taken idiosyncratically by some . . . in one case in the subject's weeping disconsolately" (p. 6).

CASE 26 was a female college student volunteer in a hypnosis experiment. She experienced "dizziness and slight nausea that persisted for 3½ hours" after taking the *Stanford Hypnotic Susceptibility Scale, Form C* (1962). "I felt I had lost my balance and was tilting. I felt I was no longer sitting upright, as though I was moving forward, floating upward. I lost my stability." During debriefing she reported that at age 12 she had a negative experience when given sodium pentothal while two teeth were extracted. She told herself: "I'm going to fight it." As for hypnosis she said: "I wanted to be hypnotized but I was experimenting. I attempted to be conscious of everything while I was being hypnotized, and more or less critical of my reaction, the same as with the sodium pentothal" (J. Hilgard, 1974, p. 293).

CASE 27 was a male college student in the same experiment as described above. He was physically active and fit, a member of the wrestling team. He "came out of hypnosis with a curiously stiff arm" and that night dreamed: "Five nuns were sitting in a row. All were crying. The woman in charge was in plain clothes instead of a habit. She was subjecting the nuns to emotional suffering. I told the woman that what she was making the nuns do was all wrong. We argued. I wasn't getting through to her. I was frustrated." J. Hilgard commented that "in this 'projective dream' S is telling us that there are substantial negative implications in hypnosis" (J. Hilgard, 1974, p. 292).

CASE 28 was a male college student volunteer in the same experiment as the two cases above. He suffered the most severe after effects. Hypnotized at 4:00 PM with the *Stanford Hypnotic Susceptibilty Scale, Form C* (1962), he was "confused, anxious, walked around in a daze, felt 'queasy.' He tried to remember things but could not. He was still dazed when he went to sleep about 10:00 or 11:00 PM; his confusion lasted throughout the night and was present the next morning for a while. He also had a headache for a couple of hours. The night after hypnosis he had many vivid dreams—they had to do with embarassment over mistakes." He described his reaction and responses to hypnosis: "I didn't come out of it. I was walking around in a daze or stupor. I didn't know what I was doing. That was scary as I look back. I remember he said to come out of it. He then started to count. I don't remember hearing him finish counting. It's really freaky—when eating dinner I wasn't thinking, just doing things automatically" (J. Hilgard, 1974, p. 289-290).

Josephine Hilgard (1974) considered ambivalent attitude and response in hypnosis to be "natural reactions to very novel experiences" such as "associations of mystery and magic due to previous contexts . . . TV and the stage." She did not consider them as a "threat to ego integration, for a normal ego is always integrating new experiences that have their plus and minus aspects. We have found very little in the way of long range threats to the ego in our experiments. The most severe case . . . was really none the worse for his experience" (p. 295).

Other factors which influence a person's attitudes and expectations of hypnosis are **folklore** (fiction books and articles, movies, TV, radio dramas), **misinformation** (from print and visual media, friends, one's own misconceptions), **regressive fears** (being dominated, overpowered, losing one's free will or moral judgment), and these can combine with others described here to make hypnosis an "emotionally charged experience" (Hilgard, 1965; Orne, 1965; Weitzenhoffer, 1957; Williams, 1953).

MISSED MESSAGES

Attentional or intellectual deficits, perceptual distortion, misunderstanding, or a subject's naivete can result in "missed messages" lessening the effectiveness of hypnosis and increasing the risk of complications. The effectiveness of hypnosis depends on the subject's ability to know, follow and understand what is being said, assuming the subject is hypnotizable. Limitations of comprehension will lessen hypnosis effectiveness. It may seem such an observation is an oversimplification but the writer has found this to be a common contributing factor to treatment failure and unwanted side effects. If a person in hypnosis perceives that some behavioral change is expected but not fully understood, the resulting frustration or confusion can be manifested as anxiety, headache, dizziness, gastric distress, or other complications.

"Ericksonian" hypnotists use **indirect methods** such as **metaphors** quite extensively (Erickson & Rossi, 1981). Proponents of this technique report a high number of treatment successes and dramatic breakthroughs in therapy. Few failures are reported. The implication is that most subjects respond well to these techniques. Milton Erickson was apparently very successful in using them. It remains to be seen whether those trained after his death, by those with limited exposure to Erickson, can consistently achieve good results. Erickson himself never founded a "school" and did not train anywhere near the number of "Ericksonian hypnotherapists" who have attended workshops since his death. Some experienced clinical and experimental hypnotists question whether Erickson's intuitive mastery of human nature and choice of the most effective intervention for an individual can be taught.

No single method of hypnosis is suited to everyone, any more than a system or method of therapy will help everyone in need. Individual differences require individualized treatment plans and methods. Not everyone has the ability to abstract, as is evident in clinical practice when subjects are asked to rephrase the meaning of such sayings as "people in glass houses shouldn't throw stones" and "rolling stones gather no moss." Many clinicians joke about their own difficulty in understanding abstract proverbs. There is no research data to demonstrate the effectiveness of therapeutic metaphor with those who cannot abstract. It seems likely that those with difficulty abstracting would have difficulty understanding therapeutic metaphors. This is an area in need of further study and until there is more data it is wise to proceed with caution.

The same is true for so-called **"right hemisphere techniques"** such as guided imagery, symbolic thought, suggested dreams or tuning the headphone on the right ear louder than the left when using autohypnosis autiotapes. Some therapists use these techniques to "offset left hemisphere dominance" ostensibly equalizing hemspheric brain function. Again, more empirical research is needed to demonstrate the efficacy of this technique. It is possible that hypnosis involves subcortical brain function such as the limbic system or brain stem. We know that the "dream center" is subcortical, in the pons, and hypnotically induced hallucinations and trance logic relate as much to dream imagery and dreamlike phenomena as cortical function. Hypnosis not only eludes precise definition but also its locus or site of function in the brain.

Paradoxical techniques can backfire if the subject takes them literally. The ridiculous extreme prescribed or suggested can embarrass and confuse the subject. The writer has had clients self-referred who have complained about "crazy suggestions" of therapists using paradox. A woman who passively tolerated the physical abuse of her husband, told she wasn't submissive enough, was beaten more. Another, told to "turn the tables" and be as or more assertive than her husband, terminated therapy when she ended up in the emergency room of the local hospital. These cases are not cited to discredit paradoxical interventions. Paradox works, but not with everyone.

Confusion techniques can unfortunately be just that, confusing subjects to such an extent they "miss the point." This technique may succeed as intended, circumvent ego defenses and overcome resistance, but so cloud consciousness as to interfere with concentration and the ability to follow a meaningful thought process, raise rather than reduce anxiety, lower the stress threshold, and diminish the effectiveness of the hypnotic process.

Deception is another technique used for the same purpose, by distraction to elude ego defenses and overcome defensiveness or resistance. It may mislead the subject by the use of words, phrases, or imagery not directly or indirectly related to treatment plans, taking the subject off on tangents with the same potential failure as confusion techniques. In addition, it raises ethical questions involving informed consent if the subject has not agreed to use of hypnosis before the deception technique begins.

Non-English speaking subjects may have difficulty understanding and responding to hypnotic suggestions. Those with a significant level of mental retardation or organic impairment from accident, stroke or a

disease process may respond poorly to hypnosis if they do not understand what is said to them. They can misunderstand the meaning of certain words, even simple terms. To those whose first language is not
English, certain terms in common usage can be confusing. Certain
words can take on added, deeper meaning because of the subject's
unique interpretation of them or if the same words were used during
painful experiences in earlier years. The hypnotist's intonation, emphasis or repetition can be misinterpreted as irritation, impatience or sar-
casm.

Some subjects may misunderstand or be misled by metaphors, paradox, confusion, or deception techniques which are intended to overcome
defensiveness or resistance, or deepen trance and elicit insight, not necessarily to be taken literally word for word. As a general rule of thumb, a
subject should never be told anything in hypnosis which if taken literally
would result in a situation for which the subject is not prepared or would
be antitherapeutic.

UNFINISHED BUSINESS

Underlying characterological problems or unremembered unpleasant events, "unfinished business" of the mind, can suddenly and
unexpectedly emerge with such force that ego defenses are swept away,
reality testing and coping skills impaired, and mental processes severely
disrupted. Josephine Hilgard (1974) reported "more prolonged reactions
after hypnosis occasionally showed indications of ego disorganization
before the hypnotic session was terminated" (p. 286).

CASE 29 was a female college student volunteer in a hypnosis experiment which included the hypnotic dream item of the *Stanford Hypnotic Susceptibility Scale, Form C* (1962). While she suffered only moderate
complications (drowsiness "lasting for an hour or so") she described what
she thought and felt during the hypnotic dream: "There was a fish swimming around. A single plane in the clouds did a nose dive and then it
crashed. It hit the ground in a clear spot in the jungle. The fish became a
plane and fell into a tail spin — all was far away and fuzzy — I never had a
sensation like that before. I was so apart from myself. My hands were
twenty feet from my body. That was a weird sensation. My mind also os-
cillated. My whole body was swinging back and forth." She was "a com-
petitive long-distance swimmer," J. Hilgard observed, "and the dream
probably had projective significance" (J. Hilgard, 1974, p. 286).

The above case gives us some indication of the nature of thought processes which can take place in the hypnotic state. Fantasy and symbolization were mixed with reality. The following clinical case describes how fantasy and symbolization which contributed to aberrant and antisocial behavior can be "unravelled" by hypnosis.

CASE 30 was a 24-year-old fireman's wife who was arrested for arson, for setting fire to her own home. There had been several fires reported in the neighborhood and, unknown to her, the fire marshal had parked in an unmarked car just across the street. He saw flames in the living room window, turned in the alarm by radio from his car, then ran into the house, seeing no one enter or leave. The woman was in the shower. When arrested she became so agitated she was judged unfit to stand trial and committed to a mental hospital for treatment. There, minor tranquilizers did not relieve her agitation and she agreed to hypnosis for tension reduction. She proved to be a good subject and her anxiety disappeared within minutes of the first induction. But she suddenly exclaimed: "I did it! My God, I did it! I set fire to my own house."

Her realization and remembering constituted a confession. She was, after all, in a state mental hospital, on arson charges, being treated to achieve legal competence for trial. If she or her attorney had known what was going to happen in hypnosis, it was very doubtful they would have given informed consent. Certainly, such a "confession" at that time may not have been in her best interests. During the trial she described her deep love for her fireman husband, who was so dedicated to his work that she felt abandoned and ignored. She got his attention the only way she knew how, by setting a fire. She was convicted but with a suspended sentence and ordered to undergo psychotherapy (Author).

In 1887, Bjornstrom described such spontaneous confessions: "Drawing forth under hypnosis confessions and secrets which they would not voluntarily disclose when awake . . . may not succeed with all somnambulists, for some are very cautious and reserved and some may even play the hypocrite, and lie and deceive their hypnotizer. But the great majority will prove very frank and outspoken, and during the sleep may much too easily hurt themselves or others by revealing secrets which ought to be kept" (1970, p. 415).

CASE 31 was a 23-year-old woman hypnotized on stage by an entertainer who suggested she "feel like a baby crying for her mother." She did so to the enjoyment of the audience and she returned to her seat. Subsequently she became increasingly depressed and required psychiatric treatment for five months. She later successfully sued the hypnotist for assault and damages (Marcuse, 1964).

Weitzenhoffer (1957) contended that age regressions are "potentially the most risky hypnosis phenomena. There is the possibility of inadvertently regressing a subject to a traumatic experience . . ." (p. 355).

CASE 32 was a male college student volunteer in a hypnosis experiment. On the age regression item of the *Stanford Hypnotic Susceptibility Scale, Form C* (1962) he reported that he relived "seeing President Kennedy's death on TV" and "felt his heart pound, and he experienced intense anxiety." After this experimental item and following hypnosis "he felt quite relaxed and comfortable"(Hilgard, 1974, p. 286).

Revivification, as Freud reported, can be as overpowering in its emotional force as the original traumatic event. War time experiences, severe stress, catastrophic events, sudden loss of loved ones, buried by time and preference, seldom remain buried as deeply as wished. In therapy they can more easily be integrated and processed since they directly relate to treatment goals. In research settings they almost always interfere with the experiment's objectives, and the traumatized subject should be debriefed and referred immediately for clinical followup. If the "traumatic insight" (Meares, 1960) occurs on stage in an entertainment setting, the problem becomes more difficult to resolve, for "help" is not as readily available. Biddle (1967) described such an experience as the "same predicament as the child awakened by a nightmare with no one present to help control the frightening dream material" (p. 6).

CASE 33 was a middle-aged man with a "classical hostile depression" who was treated with hypnosis for relief of severe headaches. In trance he was told he would not feel pain "but what it actually was and where it belonged." The headache abated but his throat "felt tight." He was given the same suggestion and this time "developed severe right abdominal pain" for which he felt surgery was required. Again the same suggestion was made and he felt "an over-full bladder" replaced by a need "to ejaculate, that he would die if he didn't." He associated these thoughts with his wife, then wept as he recalled his mother's death. "All he could think of was her convulsions" which he associated with sexual intercourse: "God! How can a man do that to a woman? He can't have his wife's death on his hands just because he wants to relieve himself" (Rosen, 1962, pp. 672-673).

SECONDARY GAIN

Meares (1961) referred to possible "perverse motivation" of some who seek to be hypnotized: ". . . it is not so much desire for the relief of the symptom . . . but rather the experience of hypnosis itself. The patient is not aware of this motivation . . . When a patient asks for hypnosis in preference to other methods of treatment we should always seek his real reason" (page 90).

In the writer's experience a few subjects—about one in a hundred—are "hypnosis addicts" or "hypnosis freaks" who go from one hypnotist to another, apparently getting much secondary gain from experiencing hypnosis. Excellent subjects, they go into deep trance from the first induction, a diagnostic sign of their obsessive drive. Other indications of this unusual preoccupation are symptom substitution to prolonged treatment, lingering in the hypnotized state, and strong transference or dependency.

CASE 34. There can be other motives for seeking hypnosis unique to individual needs. A widow, hypnotized successfully by her dentist so enjoyed and benefitted from its use, asked that he hypnotize her for every dental procedure in every treatment session. Some time after dental work was complete she phoned to ask if he would hypnotize her to help her stop smoking. He did so but she did not respond well despite several sessions. She implored him to keep trying, but this was also unsuccessful. Concerned that her compulsive need for and insistence on hypnosis might signal some underlying mental problem, he referred her to a psychiatrist (Rosen, 1960, p. 141).

In the course of psychiatric treatment it was found that she smoked only a few cigarettes a day, hardly enough to constitute a serious threat to her health. Her motive in continuing to consult the dentist emerged in the course of psychotherapy: "She was a lonely widow who hoped her dentist would fall in love with her." For her, hypnosis was the most intimate sharing she could arrange with him and she sought to prolong it as long as possible. Interestingly, Rosen observed: "Needless to say, she was not hypnotized by the psychiatrist."

There is another group of persons who actively seek the hypnotic experience. These are more "faddist" than "addictive." They collect hypnosis, relaxation, and self-help books, articles and tapes and continually build an extensive library of these materials. Both the "addicts" and the "faddists" become very knowledgeable about hypnosis but neither ever complete their quest, seeking yet another hypnotic experience by yet another hypnotist or adding another book, article or tape to an always growing but never complete personal library.

RESISTANCE

It is probably a valid generalization that everyone is resistant to someone or something at one time or another, such as toward an overbearing salesman, telephone solicitors, religious cultists, a boring speaker, unexpected company, demanding children, neighbors or relatives with unreasonable requests. Much resistance involves ambivalence, the push-pull of coexistent love and hate. We want to be accepted, to accept others, to be a friend, but it is also very difficult "to be all things to all people."

Hypnosis subjects show ambivalence when they agree to be hypnotized for some specific agreed goal or purpose then resist the hypnotist and/or the hypnotic process. Ironically, clinical subjects actually pay for the opportunity to sabotage their own therapy. Resistance to hypnosis takes many forms. It can be conscious or unconscious, direct or indirect, a failure to awaken on time as directed or premature waking, allowing interfering thoughts to prevent trance deepening, defense mechanisms unique to the subject's personality, unwanted side effects (most frequently headache, drowsiness, anxiety, pain or discomfort) or denying any effect at all ("I really didn't feel any effect at all").

Meares (1960) described several kinds of resistance which the author characterizes as: anxiety due to inability or fear of ability to carry out posthypnotic suggestion; agitation and/or confusion from consciously resisting hypnosis but unconsciously unable to do so; hostility, refusing to come out of trance; comfortably indifferent, so enjoying the imagery or serenity of trance they delay or refuse awakening. These are quite similar to Masling's "screw you" behaviors (1966) and the passive obstructionism of "the low-level employee" (Argyris, 1968). The writer has encountered quite a few usually bright subjects who intellectualize or "head trip" to avoid trance, "one upmanship" to compete with or elude the hypnotist. Others resist to punish the hypnotist for too abruptly ending their peaceful hypnotic rest, to defy a symbolized parent in a childish "No, I won't" or "See, you can't make me!" And some resist when they perceive the posthypnotic suggestion to be unacceptable or unattainable.

CASE 35. A quiet, timid student volunteered to be hypnotized in a demonstration in an introductory psychology class. He became belligerent, drew back as if to strike the hypnotist. He refused to awaken for one hour during which he wept, complaining of "how coarse" his fellow students were. Dehypnosis completed, he slept most of that day (Williams, 1953, p. 4).

CASE 36. This case demonstrates a refusal to awaken but because of anxiety and fear, not hostility. A 15-year-old high school student was unintentionally hypnotized as part of a "guided fantasy" exercise. Students seated in class were asked to close their eyes as the teacher described "a trip in space." Just after imaginary blastoff the school bell rang, a loud clanging just outside the classroom door. All students quickly opened their eyes but this young man froze in fear, clutching the edge of the desk with both hands, his eyes tightly shut. The teacher was unable to dehypnotize him. The principal was summoned and he, too, was unsuccessful. A local clinical psychologist experienced in hypnosis was called in, spoke calmly and softly and succeeded in "returning him safely to earth and the classroom" (Author).

CASE 37. Hallucination was induced in a young man during a hypnosis demonstration. When it was suggested he return to reality, the subject became irritated and pointedly told the hypnotist that he should stop talking because "he was interested in seeing the movie through to its completion" and that "he intended to remain until the end of the picture and see it through a second time." The subject sought to process the emotional reaction or meaning of the suggested experience. Dehypnosis interfered with that need and the subject resisted it (Williams, 1953, pp. 5, 6).

CASE 38. A woman successfully treated with hypnosis for "hysterical aphasia" refused in trance to reinstate the aphasia stating that "she did not wish to become aphasic on awakening" and if the hypnotist persisted "she would not allow herself to be awakened." The hypnotist did persist but without success. "Dehypnotizing was possible only when the period of aphasia was reduced to five minutes." Here, the subject had a better idea of therapeutic progress than the hypnotist-therapist. She was ready to move on (Williams, 1953, p. 5).

CASE 39. A woman convicted of murdering her husband "lapsed into a hysterical coma . . . terminated only with great difficulty after 158 hours." When finally awakened she said: "I want to go back where I was, away from staring eyes and fingers that point at me crying 'You're guilty—you must pay' " (Williams, 1953, p. 8).

CASE 40 was a young man trained and rehearsed for a hypnosis demonstration. During the actual public demonstration all went well until it was time to dehypnotize. He did not respond. Asked why he was not willing to awaken as he had before he said that he would not awaken until promised treatment for whatever "he felt his difficulty to be." Agreed, he awakened. "During treament it was revealed that the trance afforded him the only relief he could find from a homosexual anxiety verging on panic and which had led him to contemplate suicide" (Williams, 1953, p. 8).

CASE 41 was a woman who replied "I'm not going to wake up" when suggested to do so. Asked why, she responded: "I'm too comfortable." She refused to talk further, remaining in trance 15-20 minutes, then spontaneously awakening. She then explained that "she had been under much tension and it was such a relief to lose it under hypnosis that she did not want to return to reality again" (Williams, 1953, p. 8).

CASE 42 was a 50-year-old male with no prior history of referral for or treatment of mental problems. The hypnotist gave his routine suggestion that on a numerical count the subject would open his eyes "rested and refreshed." The client calmly replied in trance: "I don't feel like it. I'm not waking up." The hypnotist waited a few minutes then repeated the suggestion to awaken. "I'm too comfortable," the subject replied. The hypnotist waited again and after a few minutes more calmly repeated the suggestion to awaken. This time, and after a total of fifteen minutes expended in attempting to end the trance, the subject awakened, smiling and with no untoward effect (Author).

CASE 43 was a 42-year-old career woman who agreed to hypnosis for habit control — to stop smoking. Hypnosis was explained and her informed consent obtained. Progressive physical relaxation was suggested and she seemed calm and comfortable. A numerical countdown was begun from 30 to 1 and at 25 she suddenly opened her eyes wide and sat upright, fully alert and awake, and with a weak smile on her face. A second attempt resulted in the same wide awake response, this time at the numeral 15. She consented to a third try and this time countdown was completed 30 to 1 but at 1 she became fully awake, wide-eyed and quite unhypnotizable (Author).

CASE 44. A university student volunteer in a hypnosis experiment successfully performed ten of twelve tasks on the *Harvard Group Scale of Hypnotic Suggestibility* from a tape recorded instructions. In trance he said he felt "heavy" and "tired" and was "allowed to rest a few seconds." Followup next day when the tasks were repeated found him unable to experience the fly hallucination. On the third day he reported difficulty waking up that morning, falling asleep in class, difficulty concentrating, and on retesting was totally amnesic to the fly hallucination. Four weeks later he was debriefed for the experiment and listened to a dehypnosis tape and he was able to replicate his performance on the first day of the experiment (Sakata, 1968).

Williams (1953) observed that resistance often occurs in subsequent hypnosis sessions, not always the first, when the subject has "learned the ropes," become more aware of hypnotic technique and learned to use it as a defense. Hilgard (1974) and Hilgard, Hilgard and Newman (1961) reported that resistance can occur when there is conflict between subject personality characteristics and the demands of the experiment. Bramwell (1903, page 377), one of the early hypnosis pioneers and historians, wrote that subjects can dehypnotize themselves — awaken spontaneously — "when presented with an impossible or obnoxious task."

The following is a listing of typical causes of resistance to hypnosis:

To defy authority or an authority figure

Regression

"I can't stand anyone staring at me"

"I feel silly" (these people often giggle or laugh during induction attempts)

"I'm afraid I won't wake up"

"I'm afraid — I don't know why" (may be fear of unknown, magic, occult, Hollywood movies/TV conditioning)

Rationalization

"I really don't have the time"

"I can't afford it"

"It doesn't always work" or

"Unless you can guarantee results . . ."

"I don't think I can be hypnotized"

Fear of change

- failure
- losing control, submitting, letting go hypnotist's motives, what s/he might do physically, sexually, mentally
- subject's motives — what s/he might do/say/feel
- death
- helplessness, dependency
- revealing strong feelings

- "going crazy"
- remembering/reliving painful experiences
- lasting unwanted effects on attitude, behavior, or personality

SYMPTOM REMOVAL OR SUBSTITUTION

Symptoms are outward manifestations of inner processes much like the tips of icebergs. Hypnosis can be the doomed HMS Titanic if it proceeds with the same lack of caution, overconfidence and inappropriate speed. The symptom can be a pain state, habit, thought or feeling and should always be carefully evaluated not solely for what it is but for what may lie below, what it may mean or symbolize to the subject. Gravitz et al. (1982) observed that "the longer a physical symptom has been part of an individual's daily experience the more likely it is that the symptom has taken on psychological meaning (p. 300).

Pain. It is difficult sometimes to separate out psychological from physical factors in a pain state, the functional from the organic components. There is a synergistic effect between physical pain and emotional distress from it. Longstanding pain is annoying and uncomfortable and this adds to its intensity. "Pain can persist for emotional rather than physical reasons," Rosen wrote in 1960, and can be "either a depressive equivalent" or serve "to hold a depression on leash." He concluded that "a single symptom can often be the last defense against decompensation" and therefore "suggesting away a single symptom may be dangerous unless there are symptoms to fall back on" (p. 142).

CASE 45. Smoking for many is a relaxing pastime, reducing anxiety, despite being a health hazard. If it is the major or only means of relieving tension, removing it without replacement may increase blood pressure up to fifty points and "apparently innocuous posthypnotic suggestions . . . may impose an additional strain on an already overburdened cardiovascular system and be harmful instead of helpful. A 42-year-old man hospitalized for emphysema was badly addicted to smoking and despite his serious medical conditions he smoked two packs of cigarettes a day. Hypnosis was used for habit control. He stopped smoking but his breathing, more relaxed but also more shallow, so limited his oxygen supply it aggravated his condition (Author).

CASE 46 was a case often repeated in clinical practice. The subject was a compulsive cigarette smoker who agreed to hypnosis for control of that habit. It was successful but several months later she had gained forty pounds. She consulted a second hypnotist. She was able to diet and lose weight but then drank heavily, consulting a third hypnotist. "To her, her compulsive chain-smoking, overeating, and alcohol intake were three completely unrelated problems." Ironically, she spoke "in glowing terms" of the three hypnotists "much like quack cancer cure testimonials made by patients a few months before they are hospitalized in extremis." Three different hypnotists treated her only for the tips of icebergs (smoking, obesity, drinking) and ignored the underlying personality problem (Rosen, 1960, p. 141).

CASE 47 was a 41-year-old man treated successfully for flying or airplane phobia. He was able to fly using autohypnosis but the next day he was fearful, anxious, agitated, and depressed. He complained that he felt "let down . . . like doom was hanging over me." He suffered from tachycardia, gastric upset and loss of appetite, diarrhea and back pain. He "remained in bed, trembling and afraid to move." He obsessed over "everything he thought, said and did, searching for hidden reasons and meanings." He had an almost delusional belief that flying was wrong for him, a personal danger or threat, and he perceived his mental and physical problems after flying as punishment for wrongdoing.

He was hospitalized and the treatment plan was "to restore his personality to its status before the airplane trip" which meant no attempts to fly again. After several weeks of sedation and rest he felt he was "at a transitional point" in his life, that it was time to become more passive and less compulsive. In subsequent weeks he became indecisive, with "what if" obsessions. After a total of four months of intensive treatment he was able to return to his pretraumatic life situation but he felt that he had changed and been changed by the experience. Symptom removal "can be hazardous" and "should not be used when the symptom is related to an obsessive-compulsive system or to a serious personality, character or ego defect" (Meldman, 1960, pp. 103-105).

Spiegel (1967) suggests that therapists may project uncertainty if they are apprehensive about symptom removal: "If the therapist expects harm from symptom removal he can inadvertently convey his anxiety to a sensitive, pliant patient." In the manner of a self-fulfilling prophecy, the patient "will then comply — by not only retaining the disabling symptom, but also be developing additional reaction symptoms." Spiegel concluded that "it is the ineptness of the therapist rather than the process of symptom removal per se which leads to negative results" (pp. 1281-1282).

SUMMARY

Subjects who are hypnotized in clinical, experimental and stage entertainment settings have within themselves a variety of potential risk factors conducive to the formation of complications. Any aspect of previous life experience or present mental state can become a negative extraneous variable, interfere with the hypnotic process and endanger mental health. Typical among these are personality traits or disorders, attitude, expectations, unresolved conflict(s), traumas beyond conscious recall, significant psychosocial deprivation, a rigid moral code or value system, difficulty or inability understanding direct or indirect suggestion and inappropriate symptom removal. Resistance, secondary gain and symptom substitution are often used by subjects as unconscious defenses if hypnosis or the hypnotist are perceived as a threat to the integrity of the personality. Severe complications are likely to occur when hypnosis weakens ego defenses and coping skills and lowers the stress threshold, leaving the personality defenseless to powerful intrapsychic destructive forces.

CONCLUSIONS

Subject/patient risk factors are not always evident and it is recommended that hypnotists routinely take detailed psychosocial and medical histories to identify potential risks. Subject behaviors and responses should be carefully observed before, during and after hypnotic induction so as to intervene promptly if and when complications occur. Followup care should be provided as needed.

CHAPTER 3

HYPNOTIST RISK FACTORS

Know thyself

Inscription at the Delphic Oracle
(circa 600 B.C.)

HYPNOTIST risk factors can be personal or professional, direct or indirect, consciously perceived by the hypnotist or unconscious, unintentional, beyond conscious awareness. Whatever hypnotists say or do, whether in a clinical, research or stage entertainment setting, can become a risk factor precipitating unwanted side effects. Personal factors involve hypnotist personality dynamics, mental state, attitude, needs, verbal and nonverbal behaviors. Professional risk factors can be lack of knowledge or skill, inadequate screening or history, poor observational skills, missed cues or misdiagnosis, ambiguous or confusing suggestions, poor timing or judgment, inappropriate choice of words, imagery or intervention, ineffective dehypnosis or debriefing and lack of followup as needed.

PERSONAL RISK FACTORS

Any hypnotist with personal problems—alcohol or drug abuse, depressed, agitated, in a conflict situation or under severe stress—who is not well or not feeling well—increases the risk of creating or exaggerating unwanted side effects. Risk factors described for hypnotized subjects in the previous chapter also apply to hypnotists. There are additional traits and behaviors unique to hypnotists.

Personality Dynamics

Meares (1961) observed that many hypnotists tend to be "rather odd characters" (p. 96). He suspected that they sought to "satisfy inner drives of their own personality when they are hypnotizing others" (p. 90). Stage hypnotists tend to use an authoritative or charismatic style, a commanding presence which suggests an underlying need for power or control. Clinicians may seek through hypnosis to satisfy a morbid curiosity or vicariously experience the subject's situation, often with sexual overtones. Hypnosis researchers may be these same "inner drives." Hippocrates (c. 460-400 B.C.) described patients' sensitivity to therapists' intentions: "For some patients, though conscious that their condition is perilous, recover their health simply through their contentment with the goodness of the physician."

Hypnotist personality traits cannot only become involved in the hypnotic process but reinforced by it. Meares (1960) claimed that ". . . hysteroid practitioners have become more hysteroid as a result of the continual satisfaction of this element of their personality" and "authoritative persons have become more authoritative." He concluded that "the practice of hypnosis is a possible danger in the way that it may tend to exaggerate certain facets of his personality" (p. 96).

CASE 48 was a handsome, young clinical psychologist very much interested in hypnosis. He attended introductory and advanced hypnosis workshops over the next few years. A female patient reported him to the ethics committee of the state psychological association. The patient's charges could not be substantiated but during the investigation it was determined that most of the patients he hypnotized were women within five years of his own age. He gave his specialization as "the general practice of clinical psychology" (Author).

Verbal Behaviors

Verbal behaviors of hypnotists which can cause or contribute to complications are: tone of voice, volume, rate of speech, pauses, vocabulary and choice of words or imagery. A loud authoritative voice can be perceived by subjects as coercive, activating a defensive reaction to unresolved conflict with parental or authority figures. Biddle (1967) held that an "authoritarian type of induction invites a battle of wills" (p. 5). Conversely, a soft, sensual voice can be perceived as seductive and arouse sexual feelings or a defensive reaction against them. A voice that reflects anxiety or uncertainty can engender similar emotions in the subject.

Rosen (1962) observed: "If during hypnotic induction the hypnotist speaks in an authoritarian tone, the patient may react in one way; if his voice is soothing, sleepy, or monotonous, in another. If the hypnotist creates by his speech the type of fairy tale atmosphere that adults use with small children, something still different may be involved . . . this relationship can be an exceedingly archaic one" (p. 644). Freudian theorists might explain this "fairy tale atmosphere" as regressive, an attempt to return to the parent, perhaps ultimately to the womb. Jungian theorists might see it as a search for a primodial archetype, behaviorists as seeking security from a significant other. Whatever the theoretical interpretation, the hypnotist's verbalizations are an important variable in the hypnotic process and a potentially negative risk factor.

Hypnosis should be aware of these aspects of verbal behaviors in order to minimize negative or defensive reactions of subjects due to their own sensitivity but also because of unintended cues from the hypnotist. An audiotape of a "standard induction" can be instructive in this regard as well a soliciting feedback from colleagues or even family members.

Nonverbal Behaviors

Hypnotist nonverbal behaviors conducive to complications involve inappropriate touching, posture, distance, eye contact and clothing. Touching includes holding hands, grasping or moving arms or legs to suggest levitation or to deepen trance or the hypnotist placing a hand on the subject, usually the forehead or shoulder. Where and how the subject is touched can influence the subject's perception.

A hypnotist who sits too far from the subject may be perceived as detached and uninterested. Folded arms, frequently looking out the window or away from the subject, can increase psychological distance. Sitting too close, leaning into the subject, with noses almost touching, is likely to be perceived by subjects as confrontive and exploitive. A female hypnotist with open blouse and no bra or a hairy-chested male with open shirt can by their attire interfere with rapport and lessen the effectiveness of the hypnotic process. If this is the only variation it is of minimal risk but if other negative factors occur, the overall risk factor increases. It is as if a critical mass is reached, fomenting unwanted side effects.

Sexual Factors

These can be conscious or unconscious, direct or indirect. Most often they are unconscious and indirect. A hypnotist who finds a subject sexually attractive is already in a risk situation. Frequently, signs of potential sensitivity are evident when taking a good history (e.g., unhappy marriage, recent divorce or ended relationship, promiscuity, generalized anxiety or high stress). In some cases there is no apparent susceptibility or sexual attraction but the subject responds erotically anyway. Meares (1961) suggests hypnotists review their clientele to compare the number of persons of the opposite sex to assess possible "latent erotic drive or psychosexual inversion" (p. 91). Sexual involvement, real or imagined, is a leading cause of malpractice litigation and hypnotists should proceed with caution, exerting constant awareness of possible emotional or sexual involvement with a subject.

CASE 49 was a 34-year-old female inpatient in a mental hospital diagnosed and treated for depression and with borderline traits. She had been sexually abused by a family member when she was nine years old. Hypnosis was used for relaxation and as an adjunct to psychotherapy. The male therapist arranged for a female staff member to be present at the twice weekly hypnosis sessions but due to staff shortages this was not always possible. One year after she was discharged she attempted to sue the therapist, the hospital and the state for "sexual advances including sexual intercourse while hypnotized." When it became known from staff and former patients that she had made statements while hospitalized she was "going to get a free ride from the state" plans for filing suit were dropped (Author).

PROFESSIONAL RISK FACTORS

Lack of knowledge and skills are major professional risk factors of hypnotists. These are due to poor training, inadequate supervision or inexperience. They are manifested in a lack of awareness of risk situations and risk factors and poor technique, by error or omission. Each hypnosis setting has additional risk factors: clinicians are usually alone with the client and are limited by their own knowledge and skills or lack of them; researchers lack clinical skills to recognize and promptly intervene to prevent complications; stage entertainers lack clinical skills and hypnotize under the highest risk conditions (e.g., poor prescreening, large numbers of subjects on stage and in the audience difficult to observe closely and individually, superficial debriefing and lack of follow-up).

CASE 50 was a 24-year-old female treated by a lay hypnotist for anxiety, insomina and restlessness. One night she became so agitated she phoned a friend who came to her apartment and drove her to the local hospital emergency room. She had been treated one year by the lay hypnotist in weekly sessions augmented by audiotape cassettes for autohypnosis. Diagnosis: bipolar affective disorder, manic. She was transferred to a mental hospital, stabilized on medication, and discharged, her hypomanic behaviors relieved. Had she been referred to a mental health professional before, her year-long struggle would have been avoided (Author).

CASE 51 was a 20-year-old male university student who was hypnotized by a history professor who "dabbled in hypnosis." Suggestions to relax were met with paradoxical reactions of tremoring, irregular breathing and perspiration. The professor continued to suggest relaxation, repeating the same suggestions again and again, until the young man "couldn't take it any more," jumped out of the chair, out the door, and ran to his room. Unable to sleep that night he was referred to a psychiatrist whose diagnosis was unresolved grief over the death of his father four months before. The professor did not know of this recent loss nor did he make any effort to obtain a history (Author).

Lack of Screening; Poor History

As has been stated often "there is no substitute for a good history." Or put another way, "to be forewarned is to be forearmed." Many hypnosis complications could have been avoided or promptly recognized and referred for treatment and the hypnotist obtained even a brief medical or psychosocial history. Stage entertainers hypnotize strangers from the audience, knowing nothing about them. Researchers obtain more information pertinent to the experiment, less about potential risk factors. Some clinicians are "too busy" to take a history or even to review one taken by someone else. Lay hypnotist and police investigators usually function at what might be called a "technician level" and lack the training and judgment of psychiatrists or psychologists.

CASE 52 was a 44-year-old woman with heart disease and emphysema who went to a "smoking clinic" where she was hypnotized in a large room with 75 others. The hypnotist arrived from a distant city, conducted three inductions, provided everyone with an audiocassette, then left. She was not screened, observed only as one of 75 persons in the large room, and no followup was provided (Author).

Poor Observational Skills

There are many early symptoms of complications and they serve as an "early warning system" to those who are trained or who train themselves to see them. In hypnosis, the face is one of repose, the facial nerves and muscles being in a relaxed state. There should be no tremoring of the lip, squinting, or eyelid flutter though many hypnotists feel that eyelid flutter can be a reflex firing of nerves due ot relaxation rather than anxiety. The same is true for long muscles, most notably in the thigh, which can "fire" and cause a twitching when the person is deeply hypnotized or deeply relaxed.

In hypnosis, the jaw relaxes so that upper and lower teeth do not touch and often the mouth is partly open. Hands if on the subject's lap are usually open, with fingers in a restful, natural position. If legs are not crossed, and many hypnotists feel they should never be, the feet rotate somewhat so that the angle from heels to toes is greater than when the person is unhypnotized. Voice quality, in those who have experienced moderate or deep trance, changes to a deeper tone and frequently subjects will clear their throat much like when awakening in the morning, due to throat muscle relaxing. Breathing rate is regular and relaxed, much like a person asleep. Heart rate is regular. Swallowing rate, if evident at all, is reduced. Ideomotor responses such as finger movement are slowed. Some hypnotists have reported changes in fingernail color intensity.

It is unlikely a subject would exhibit all these nonverbal behaviors but if none are present or there are opposite signs, the hypnotist should observe more closely, assess the situation, and take prompt action by changing the induction process, slowing it, or consider immediate dehypnosis and talk further with the subject to reduce risk of problems. With subjects who do not "look hypnotized" there may be no other way to proceed but to complete one induction and then evaluate. From the risk prevention standpoint it is important to "stop, look and listen" if a subject does not visibly respond to hypnosis.

Missed Cues; Misdiagnosis

For nonmedical hypnotists, a source of justifiable concern centers on a missed diagnosis of a serious medical problem. Physicians fear missing signs of severe mental disorder — mostly a suicidal depression. Dentists are apprehensive about both medical and mental problems. Walker

(1967) recommends taking a medical history and one is provided as an appendix to this book. While Walker spoke mainly of medical problems which can masquerade or be misdiagnosed as mental disorders, his observations apply as well to medical problems treated solely with hypnosis: "There are frequent (often tragic) instances in which patients with serious medical or neurological diseases are treated for some time solely by psychotherapy or psychoanalysis" (p. 3).

It is true that anyone can make mistakes. It is also true that it is difficult in clinical practice to separate out organic from functional causes of disorders, for a researcher to know much about either, or a stage entertainer to even understand this question. It may be true that a hypnotist's chances of being sued are probably small. But Walker does not so easily accept mistakes as inevitable: "This does not return the patient with the brain tumor from the dead or turn back the clock for the patient with advanced metabolic disease that had been reversible earlier" (p. 4). Taking a medical history is a very small price to pay to avoid litigation, perhaps to save a life.

There are certain physical symptoms which should automatically signal alarm to every hypnotist: headaches, dizziness or disorientation, poor memory and drowsiness. There are, of course, many other symptoms of medical disorders apparent in the course of taking a history or reported by subjects. Walker points out that they could signal an underlying cardiovascular disease, endocrine disorder, a neoplasm or hematoma. Impotence (secondary) can be caused by neuroleptic medication, Addison's disease, diabetes, cirrhosis, multiple sclerosis, arteriosclerosis, brain or cord tumor, or lead poisoning (Walker, 1967, pp. 41-42).

CASE 53 was a middle-aged woman with mild, infrequent seizures which occurred only when she was asleep. Medical and neurological tests were negative. She was treated with hypnosis and it was found that seizures could be precipitated by direct hypnotic suggestion. An electroencephalogram was taken during hypnosis and showed "minimal intracerebral pathology." The hypnotherapist felt that hypnosis somehow "channelized her anger." She was found to have a meningioma which was surgically removed (Rosen, 1962, p. 669).

CASE 54 was a 21-year-old woman who developed grand mal seizures a few years after an auto accident. In time they became more frequent to 35 a day. She had a "right foot drop and a mild right central facial palsy." Between seizures she was hypnotized and it was suggested the motor activity stop and that whatever "they kept her from feeling or doing she could not feel or do." She then developed "coital movements" hallucinating a former sexual partner "but she assumed the male role holding her index fingers where a penis would be, verbalizing 'obscene threats.' " Grand mal seizures decreased, disappearing in three to four weeks and they did not recur. "Although there were neurological signs, her attacks were on a psychogenic basis" (Rosen, 1962, p. 669).

These two cases demonstrate how physical symptoms can mask underlying organic or functional disorders difficult to differentiate. They point up the need for nonmedical hypnotists to take a careful medical history, inquire about physical health and the recency of medical examination, and to refer to a physician if there is any possibility of organic etiology. In like manner, medical practitioners should have a practical working knowledge of psychodiagnostics and psychopathology.

CASE 55 was a 20-year-old man treated by a psychologist for anxiety, depression, anergia and anorexia. Hypnosis relieved the anxiety but he continued to feel "out of it, beat, washed out." Actual diagnosis: mononucleosis (Author).

CASE 56 was a 39-year-old female who complained of being irritable, apathetic, restless, and with frequent severe headaches and impaired memory. She had many upper respiratory infections in the past and most recently an inner ear infection. She was treated unsuccessfully by a lay hypnotist. Actual diagnosis: meningitis (Author).

CASE 57 was a 32-year-old male with a long history of irritability and difficulties controlling his anger. In time he experienced dizziness and vomiting, with headaches. Medical examination was negative. He attended group therapy to help him control his temper augmented with autohypnosis instruction. Actual diagnosis: Intracranial neoplasm in the right temporal lobe (Author).

Ambiguous or Confusing Suggestions

Unclear, confusing or inappropriate suggestions are like giving someone half a road map and expecting them to find their way. Most likely behaviors would be for them to wander aimlessly, remain passively in the same place, or become irritated and frustrated. The same is true psychologically when subjects are given ambiguous or confusing suggestions. Some subjects are regressed to childhood years may no longer know the hypnotist. After all, the hypnotist and subject haven't met yet! Preventive practice would be for the hypnotist to identify as someone known to the subject at that age or a friend of a known and esteemed person to the subject during the regressed years. Also, it is good preventive practice to build in exits for smooth re-entry to the present, such as being in a movie (that can change quickly), a time machine (that always works) fitting the "magic" to the subject's personality. Subjects are accessible from a variety of attitudinal and motivational pathways, most frequently power, permission, prestige, novelty, intelligence or belonging.

A hypnotist suggested subjects use a deep breath when they needed it. The suggestion was: "You will use the deep breath every day and every night, as many times as you think of it, until it becomes second nature to you." **CASE 58** used it once or twice a day, reporting: "That's when I thought of it." **CASE 59** used the deep breath excessively and hyperventilated: "I needed it a lot." **CASE 60** stopped it completely, explaining that "to do it I stopped everything, closed my eyes and really concentrated on relaxing." His boss didn't appreciate him sitting at his desk with his eyes closed. Preventive practice would be to chat with the subject after each induction to ensure a reasonable understanding of posthypnotic suggestion (Author).

Inappropriate Time or Timing

Risk factors of time or timing are due to extremes. Proceed too slowly and some subjects lose interest by wandering off the subject or falling asleep. Proceed too quickly and some subjects react with a defensive irritation or anger or passively, experiencing increasing anxiety. Inexperienced hypnotists tend to move either too fast or too slowly, rarely with moderate or optimal speed. Speak at a faster rate than the subject thinks (or speaks) and the risk of anxiety or irritation increase. Plunge too deeply too soon into deep or emotionally charged material and the risk increases for fear, panic or agitation. The current emphasis on brief or time-limited therapy and limited number of sessions dictated by third party payors contributes to a "therapist in a hurry" syndrome. Early signs of developing complications can be missed "in the rush" and the hypnotist's impatience or nervousness can be projected onto the subject.

Good preventive practice is for the hypnotist to slow down rate of speech and depth of processing to keep pace with the subject and in this way keep pace with the subject's stream of consciousness. This is the key to good processing, known to skilled therapists of every generation. Janet (1925) wrote: "As a general rule the awakening should be postponed if a morbid symptom of any sort should intervene during the hypnotic state" (p. 9). On the other hand, giving total freedom in trance to subjects who are disturbed or with much powerful but repressed material can be risky as Biddle (1967) points out: "It is unwise to allow uncontrolled free association in the trance . . . it permits the patient to uncover repressed material too rapidly and usually results in danger acting out" (p. 9). Knowing when to increase or decrease speed and depth is a skill which comes with experience. Risk of complications increases as time and timing skills decrease, pointing up the need for a greater level of expertise among hypnotists.

Inappropriate Word Choice or Vocabulary

Using "the wrong word at the wrong time" can unintentionally release repressed material flooding into consciousness in a crisis situation. Even the simplest words can be emotionally charged if they are associated in the subject's mind with an earlier trauma or present conflict situation or severe stressor. "Sleep" to a person with a phobic fear of death can associate sleep with death and trigger a dramatic anxiety reaction. "Heavy" can imply "being stuck" or "sinking" and "light" to some can be associated with "lightheaded" or being "fragile" or "delicate." And "warm" to a person once burned can trigger recall of that event:

CASE 61 was a female volunteer in an experiment on hypnotic hyperthermia. She experienced spontaneous age regression to a traumatic childhood incident in which she was burned (Author).

CASE 62 was a 49-year-old salesman who consulted a clinical psychologist for hypnosis to stop smoking. During history-taking when asked if he had ever been hypnotized he described how at age 18 while in college he volunteered to be hypnotized by a philosophy professor "who used hypnosis as a hobby." Twenty-five years later he chanced to meet a former classmate and while "talking old times" recalled the impromptu hypnosis demonstration. He was fascinated to learn he was stuck with a pin which he did not feel and did not recall later. He became quite concerned when he heard he was told he would experience "seven years of feast, seven years of famine, then an average life."

He recalled that after graduating he did well in progressively better jobs: "It was as if I couldn't fail," he said. For the next seven years he failed more than he succeeded: "I couldn't seem to do anything right," he commented. "Then things seemed to settle down," he said, "I did fairly well but couldn't reach full potential." He wanted to go into business for himself "but something deep inside prevented it." He was hypnotized to help stop smoking and posthypnotic suggestion was added to reassure him of the future and that whatever happened in the past was gone and need not bother him. He was a good subject and responded well. After three sessions of three inductions each he reported "feeling free, really free, like a heavy weight has come off me." Six months later he was doing well in business for himself and not smoking.

Admittedly, this man's belief that he was hypnotized at age 18 to suffer 21 years of famine, feast and mediocrity can be explained away as the sensitivity of a young man in an identity crisis, or feelings of inadequacy and inferiority, rationalizing his own inability to succeed, or perhaps undue attention to detail of an obsessive or compulsive personality. The fact is that he was hypnotized and was a good subject, witnessed by the classmate who remembered what was said to him in trance. This case is cited as an example of the possible long term effect of what is said but more importantly as an example of irresponsible, downright stupid posthypnotic suggestion (Author).

Commercial and some professional **hypnosis audiotapes** contain words and phrases which involve some risk to users. One tape anticipates eyelid flutter and suggests: "You may notice some eyelid flutter. If you do, it won't bother you." Not everyone who is hypnotized experiences eyelid flutter. Since hypnosis involves suggestion, the mere mention of such a side effect is likely to cause it in susceptible subjects. "Your eyes are heavy. It feels like there's lead paste on them." To some "paste" is not easily removed, even with soap and water, and "lead paste" seems more difficult to remove as well as possible association with lead poisoning. These interpretations may seem hypercritical but why use verbalizations that can be misinterpreted and are not really necessary to induction? Gravitz et al. (1984) urge caution with respect to audiotapes for self-hypnosis: "Due diligence must be exercised to ensure that the client is capable of assuming the responsibilities of being both hypnotherapist and client simultaneously" (p. 309).

Other word misuse which can interfere with the hypnotic process is addressing subjects by the wrong name, or using a different induction method than the previous session. Another hindrance to induction is the overuse of words. Many therapists and researchers overuse a specialized vocabulary, technical jargon, which distances the subject from the hypnotist, placing an obstacle between them. Subjects can perceive such verbose hypnotists as haughty or condescending. The hypnotist's rate of speech or use of pauses can be perceived as impatience, irritation or indifference. These are all potentially negative factors which can interfere with affective trance induction.

Inappropriate Guided/Visual Imagery

Inappropriate imagery carries the same risk as poor word choice. In many "smoking clinics" where many people are hypnotized at one time it is not possible to select visual imagery to suit every individual present. Only one scene is used, usually an ocean beach. This works well if an ocean beach is a pleasant scene for the subject. The same is true for a mountain scene, a meadow, on a river bank, or anywhere else. The best preventive practice is to ask the subject what is a peaceful, restful scene. Even this is not without danger. There are obvious problems when the hypnotist forgets which imagery was used in previous sessions and inadvertently uses the wrong scene, one that is totally foreign to the subject.

CASE 63 was a 44-year-old woman treated with hypnosis for help in dieting. She chose an ocean resort area which she found restful but at the third hypnosis session she asked that it be changed: "I'm never alone at the beach. We go there every year. I'm there with the kids, always on the go." Good preventive practice would have been to ask her where she could go alone, a real or imagined place, not necessarily a place where she went with others (Author).

CASE 64 was a 50-year-old man who went to a smoking clinic where the ocean beach was the only visual imagery used. He had almost drowned as a child in the ocean, pulled out to sea by the waves and current. He felt that his need to stop smoking was greater than the anxiety from the childhood trauma so he tried to cooperate in the smoking clinic. He was not able to stop smoking and his sleep was troubled by dreams, reliving his near drowning. He consulted a psychologist experienced in hypnosis, neutral imagery was substituted, the dreams ceased and he was able to stop smoking (Author).

CASE 65 was a 42-year-old man who was hypnotized to curtail smoking. He chose an ocean beach for visual imagery but after the second induction complained his "ex-wife showed up!" He asked that the scene be changed: "It isn't good there anymore. That's where we went every year. We were always fighting" (Author).

Poor Judgment: Inappropriate Intervention

As was demonstrated in Chapter 1, subjects can bring with them a great variety of mental "unfinished business" which, if undetected can lead to inadvertent "uncovering" in hypnosis. In such cases, neither the hypnotist nor hypnosis are the cause of the underlying problem or disorder. They are unfortunately present at "the eleventh hour" and without evidence to the contrary play a role in the uncovering. Good preventive practice is to sharpen awareness by review of the subject's history and careful observation to prevent untimely venting or to channel and control it if it "runs" beyond control.

Meares (1961) considered unintentional, unpredictable venting during hypnosis "an acute medical emergency." He recommended hypnotists be prepared and competent to use more than one method to meet whatever emergency arises. He suggested relaxation techniques and induced amnesia posthypnotically of revivification if in the hypnotist's judgment recall would be antitherapeutic. It is a judgment call which involves the knowledge, skill and experience of the hypnotist.

CASE 66 was a young man hypnotized by a male peer "to help him succeed" in an examination. He then had bizarre and grandiose ideation, lethargy, social withdrawal and complained that he felt like he was becoming a robot." His friend dehypnotized him but the systems not only persisted but became more severe. His symptoms were relieved when he was referred to and treated by a mental health professional experienced in hypnosis who concluded the symptoms were "entirely a product of a continuing hypnotic state" (Kleinhauz & Beran, 1984, p. 287).

CASE 67 was a middle-aged female who was hypnotized on stage by an entertainer, regressed to age 11, and "began to speak in her native language." Someone in the audience who knew that language came on stage to translate. It was suggested she "act like a child playing." After the show she felt upset and consulted the hypnotist who told her to "go home and sleep." She became irritable and regressed in her behaviors to a childish level. She ignored water faucets looking for water pumps from her childhood. She experienced depersonalization and derealization for periods of a few hours to several days. Her perception was distorted, complaining she felt like she was "walking on air." This condition persisted for ten years before being treated and relieved by a psychiatrist experienced with hypnosis (Kleinhauz & Beran, 1984, pp. 285-286).

Ineffective Dehypnosis or Debriefing

Not all subjects emerge from hypnosis feeling alert and awake, even though hypnotists suggest that they do so. They may open their eyes when told to do so, and, interestingly, this varies with the individual, but their reflexes can be slowed to the degree that it is dangerous for them to drive or to return to a work setting where they operate equipment or machinery.

CASE 68 was a 44-year-old woman hypnotized to facilitate individual psychotherapy by her therapist, a clinical psychologist. Hypnosis was successful. She opened her eyes at the end of the induction as requested but said: "Oh, could I just rest a minute. I feel relaxed and slowed down and I don't feel I should drive my car feeling like this" (Author).

Good preventive practice would be to have subjects remain in the waiting room a half hour after hypnosis or until the hypnotist determines there is little likelihood of posthypnotic complications. Gravitz et al. (1982) recommended that "precautions should be taken to avoid the premature departure of a subject after hypnosis" (p. 306).

CASE 69 was a man who was experiencing paranoid ideation and obsessions that someone was following him. He would constantly look over his shoulder and he was experiencing fearfulness and anxiety. He sought professional help and it was determined that onset coincided with being hypnotized on stage by an entertainer who suggested to him in trance that a fierce dog was after him (Marcuse, 1964).

CASE 70 was a 19-year-old girl hypnotized on stage and told she would "fall asleep" whenever she heard a certain song. She collapsed at work the next day and experienced spontaneous trance states for two weeks until rehypnotized. The symptoms did not subsequently recur (Marcuse, 1964).

CASE 71 was an anonymous, public hypnotization in a televised interview during which hypnosis was demonstrated using a handclasp induction technique, suggesting that the interviewer's hands could not be unclasped. Twenty phone calls were received from friends or family members of viewers unable to unclasp their hands (H. Clagett Harding, 1977). Hypnosis induction has been broadcast by radio and while dehypnosis was included it would seem especially hazardous for highly susceptible subjects to hear an induction on their car radio while driving (Author).

One of the most serious after effects of hypnosis, reflected in several dramatic cases in this book is **spontaneous trance.** These have been reported in clinical, research and stage entertainment settings and those which have been promptly reported and treated have been resolved through rehypnosis. This would seem to implicate the hypnotist or the technique used and very likely is a combination of the hypnotist's lack of skill or judgment and failure to fully dehypnotize.

CASE 72 was a woman treated with hypnosis for phobic desensitization of a fear of elevators — she had been trapped in one for several hours. The hypnotist did not debrief her to the white sound he used during trance and when she walked past a ventilator which sounded similar she "passed out." She was not aware of the association and the hypnotist learned of it when discussing her progress and events in her life between sessions. Posthypnotic suggestion that she would respond only to white sound only during hypnosis was included in subsequent treatment sessions. The phenomenon did not recur. She had no prior history of it and she was otherwise in good mental and physical health (Author).

While the vast majority of subjects who hear white or other sound during hypnosis do not have such a dramatic reaction to it, good preventive practice would be to ensure the dehypnosis of every subject. In this way even the small percentage of such incidents would be further reduced if not totally eliminated. As for awakening procedures, a 5-count would be better than a 3-count or simply suggesting: "Now you are awake, wake up, wake up" or some similar sudden arousal. Meares (1961) observed: ". . . to make a split second change from hypnosis to waking on the count of a certain number . . . is manifestly not so." Preventive practice would be to make some other sound when saying the last number, such as switching off the white sound amplifier or a soft pat of the hypnotist's open hand on his/her own knee with a pause for "re-entry" before the hypnotist says anything. If this is unsuccessful and the subject remains hypnotized, the re-entry process should be repeated calmly and the subject given even more time to awaken. It is very rare that a third sequence is necessary but, if it is, the suggestion should be made that the subject will awaken "soon" feeling well and more time allowed. The situation should be considered a crisis only after three unsuccessful attempts.

Williams (1953) observed that incomplete dehypnotizing occurs when moods and emotions elicited in the trance continue after the terminating signal has been responded to. In others there are the familiar headaches, drowsiness and lethargy which frequently occur" (p. 6). Other factors contributing to inadequate dehypnosis are "errors and ambiguities in suggestions, omission of some detail when given the signal to terminate or when the signal to terminate the trance is complicated"(p. 7). Etiology is not so easily identified: "Partial dehypnotization or refusal to dehypnotize may arise from a combination of motives in any given case and also there are a very large number of highly individualized motives that can evoke these conditions" (p. 10).

Rehypnosis returns the subject to as nearly the same mental state as when the persisting posthypnotic suggestion was given so that it can be effectively neutralized. Any underlying psychodynamics which may have contributed to the complication can then be more easily seen and a decision made to investigate these further or give reassurance and probe no further. This is a judgment call which points up the need for hypnotists to have considerable clinical sophistication.

Lack of Followup

Many hypnotists' busy schedule makes it difficult to follow up cases after agreed treatment goals have been met. This presents little risk unless treatment has not been completed or a subject does not keep appointments and "drops out" of treatment. If a subject sustains what can be legally construed as an "injury" or "legal damages" and somehow connected with the interrupted treatment regimen, the hypnotist can be sued for malpractice:

CASE 73 was a 44-year-old man referred for individual psychotherapy by an internist. The physician referred him for habit control (excessive cigarette smoking) and tension reduction. Treatment was begun and he responded well to posthypnotic suggestion to stop smoking and to use an occasional deep breath for relaxation. He failed to keep the third appointment. No followup calls were made to him, nor any written reminders mailed. The internist did not make any inquiries of the hypnotist. Six months later the man died of a brain tumor and both the internist and hypnotist-psychologist were sued for malpractice (negligence and abandonment). Their insurance companies opted to settle out of court (Author).

E-BIAS, DEMAND CHARACTERISTICS

Hypnotists' verbal and nonverbal behaviors and the hypnosis situation itself can exert direct and indirect influence on subjects beyond the agreed purpose of the session, in treatment or research settings or on stage. Research studies in social psychology refer to these influences as **E-bias** (experimenter bias) and **demand characteristics.** Actually, James Braid aptly described them a hundred years ago as three experimental errors: (1) where "the operator and subject may both voluntarily try to deceive the spectator;" (2) where the "operator may be honest" and "the subject may try to deceive him;" (3) where "the accuracy of the experiments may be destroyed by unintentional errors on the part of the operator, the subject, or both" (Bramwell, 1956, pp. 145-146). Braid also reported "sources of error" where "the docility and sympathy of the subjects . . . tended to make them imitate the actions of others" and "deductions rapidly drawn by the subject from unintentional suggestions given by the operator" (Ibid., p. 144). These ideas were further developed by Rosenzweig (1933), Rosenthal (1866) and Orne (1962 a, b).

Rosenzweig's Three Errors

Rosenzweig's 1933 work is as timely today as when he wrote it more than a half century ago. He described three sources of error which he felt transformed an experiment into a psychological problem:

1. Errors of observational attitude. In the early years of astronomical research there were embarrassing situations where observers varied in timing the transit of planets and stars, and even such pioneer experimental psychologists as Wundt and Titchener reported differences among observers in the most rigorous controlled experiments (Boring, 1950). Rounding off fractions or decimals in a direction which favors the hypnothesis would be an error of observational attitude.

2. Errors of motivational attitude arise from the subject's awareness of the experiment's hypothesis, what it seeks to prove or disprove. Ideally, subjects should not be aware of the hypothesis because knowing it renders the subject biased and this awareness can affect the subject's responses. If a subject wrongly perceives the hypothesis this can further contaminate the experiment because responses may not be relevant to the experimental situation.

3. Errors of personality influence. These involve the experimenter's personal and physical traits, seldom obvious and rarely reported in published research. They include the experimenter's physical appearance, verbal and nonverbal behaviors which are not necessary to and interfere in the experimental process such as "an unguarded word or glance" (p. 352).

Experimenter Effects

Rosenthal (1966) saw E-bias as being essentially direct or indirect. It is indirect when it involves experimenter errors of observing, interpreting or reporting, a "halo effect," favoring the experimental hypothesis.

Direct E-bias consists of two different kinds of experimenter traits or attitudes:

1. Experimenter effects, reflections of the experimenter's personality dynamics, factors which add to or interfere with the intent of the experiment. A male psychologist, hypnotizing a female subject he finds attractive (or vice versa), is likely to involve experimenter effect.

2. Experimenter expectancy effects are the experimenter's verbal and nonverbal behaviors used in conducting the experiment which unintentionally shape responses. Approving nods or looks, pauses after certain responses, and verbal reinforcers such as "good" or "yes" are examples of expectancy effects not actually required to conduct the experiment but which can influence subject responses.

Demand Characteristics

Orne (1962b) referred to factors in an experiment which shape subject responses or trigger or increase subject awareness of the experiment's purpose as **demand characteristics.** They are the totality of cues," Orne wrote, "which convey to a subject the purpose of an experiment." They are "not merely inherent in the instructions" but can be "conveyed by the experimental design itself. By what the subject is asked to do he may . . . understand what the investigation expects him to do" and "implies we are dealing with social perceptions, and like all perceptions these are dependent upon the past experience of the perceiver" (p. 677). Orne pointed out that "different subjects may perceive different demand characteristics given the same experimental situation . . . a college undergraduate population may differ from the National Guard" (p. 678).

Throughout the history of hypnosis, hypnotist and subject behaviors have varied a great deal because it is a highly individualized, personal experience. Anything said or done can shape hypnotic and posthypnotic behaviors, and, in so doing, become demand characteristics. "If specific behavior becomes widely publicized as characteristics of hypnosis," Orne observed, "subject and hypnotist alike may view it as typical. It will then tend to occur when hypnosis is induced" (1962b, p. 681). Mesmer performed hypnosis dramatically and forcefully and his subjects responded in the same way. Charcot saw and reported many hysterical symptoms and was led to believe hypnosis was a form of hysteria. Puysegur found his subjects to be passive and relaxed. Bernheim saw hypnosis only as heightened suggestibility. Each used hypnosis in his own unique way and the majority of hypnotized subjects followed the demand characteristics projected by the hypnotist.

E-bias and demand characteristics occur in clinical, experimental and stage entertainment settings. Statements used to hypnotize can be misinterpreted with little or no opportunity to ask questions or clarify meaning. Told "side effects are rare" may reassure some (if rare it's unlikely to happen right now) or anxiety provoking (so it can go wrong and if so why not right now to me?). Told "you feel relaxed, warm and comfortable" might be perceived as having sexual overtones. Hypnotists might in their delivery overstate suggestions, perceived as overbearing and dictatorial. E-bias and demand characteristics can and do enter into the hypnotic process and anyone using hypnosis should be aware of this and seek to minimize their effect.

The history and evolution of experimenter effect parallels the developing concern about hypnosis complications and misuse. Problems coincident with or resulting from the use of hypnosis have accumulated in the clinical and research literature and in the lay press and more cases have been referred to ethics committees. Some cases have involved life threatening emergencies (Kleinhauz & Beran, 1981) and raise serious questions as to the risk potential of hypnosis. As no researcher protocol should be considered without a thorough knowledge of and review of the literature on E-bias and demand characteristics, so no hypnotist should proceed without a thorough awareness of the hypnotist, subject and environmental risk factors involved.

FORENSIC HYPNOSIS

What E-bias and demand characteristics studies have done to increase the sophistication of researchers, state courts have done for forensic hypnosis. Recent clinical case studies and empirical research conclusively prove hypnotically-enhanced memory of witness, victim or defendant can be inaccurate and distorted. James Braid prophesied this a hundred years ago when he described "the vivid state of the imagination in hypnosis which instantly invests every suggestion, idea or remembrance of past impressions with the attributes of present realities" and also hypnotist or experimenter effect, "the tendency of the human mind, in those with a great love of the marvelous, erroneously to interpret the subject's replies in accordance with their own desires" (Bramwell, 1956, p. 144).

Until the late 1970s, "hypnotically refreshed memory" was used widely in court with crime victims, witnesses, plaintiffs, to obtain confessions and to identify and present mitigating circumstances more favorable to defendants. More specifically, age or time regression techniques were used to enhance recall of simple facts (like license numbers), incidents (identify felons), and details of complex situations (scene of the crime, sequence of events). Police departments sent law enforcement officers to be trained as hypnotists, to have an immediately available pool of in-house hypnotists to facilitate recall of victims, witnesses and consenting defendants seeking to prove their innocence. In those days, hypnosis could "do no harm" and subjects were told they were able to access and accurately describe the objective reality of whatever they saw, said or did, like a camera or videotape recorder. It was not unlike those who today claim that hypnosis is safe and without danger.

CASE 74 was referred by a local rape crisis center staff. The victim was taken by three male police officers to a private residence and there hypnotized to recall her recent sexual attack. No female police officer or female advocate was present. One police officer later dated the victim! The session was not taped. The rape crisis center staff consulted a local psychologist experienced in hypnosis. He visited the police chief together with a psychiatrist and another psychologist to share their concerns. The police chief defended the competence and ethics of his officers and saw no need to take corrective action (Author).

This case occurred in 1978 and is not likely to recur because of subsequent court action which restricts the use of hypnosis. As E-bias and demand characteristics studies made researchers more aware of unintentional contamination of their experiments courts have become more aware of "what can go wrong" with hypnosis and have changed the role and practice of hypnosis in the courts. That it had to take court action state by state for more than a decade suggests that hypnotists were either not aware of the risks or were unable to join together to do anything about it.

Harding v State, 246 A2d 302 (Md 1968)

The role of hypnosis in courtroom testimony has changed dramatically since 1968 when the Maryland court in Harding v State accepted hypnotically enhanced testimony as admissible evidence. This stimulated the use of hypnosis by lay persons. For several years hypnosis was hailed as a valuable investigative tool. "Spectacular successes" were attributed to the use of hypnosis (Applebaum, 1984; Smith, 1983). Subsequent studies clearly demonstrated that hypnotic recall is not always accurate (Dywan, Hamilton & Orias, 1983; Hilgard & Loftus, 1979; Laurence & Perry, 1983; Orne, 1979, 1986; Putnam, 1979; Smith, 1983; Zelig & Beidleman, 1981).

We know now what we should have known long before, that hypnotized person's recall is seldom perfect, subject to all the variations described in Chapter 1. We know that the wording of a question in hypnosis can significantly distort recall (Hilgard & Loftus, 1979) and more errors occur in response to leading questions (Putnam, 1979). Memories from previous life events, altered by fear, fantasy, or value judgments, can contaminate the recollection of recent events (Orne, 1986). Hypnotized subjects can distort, fabricate, confabulate or deliberately lie and trained, experienced experts cannot with certainty detect it (Applebaum, 1984). After hypnotic recall this new memory is fixed permanently as historical reality and there is no way to separate fact from fantasy.

State v Hurd, 432 A2d 86 (NJ 1981)

In 1981, the New Jersey Supreme Court imposed specific limitations to the use of hypnosis in courtroom testimony. In outline form it required that to be admissible hypnosis should be done by:

> a psychiatrist or psychologist who is trained and experienced in hypnosis and not representing the prosecution and is given written instructions and also the witness' prior testimony.

The foregoing occur before the hypnosis session. The hypnosis session:

> begins with witness' conscious, unhypnotized detailed recall, then hypnotized detailed but minimally guided or led recall with no one else present and the entire session is taped.

This case marked a turning point for forensic hypnosis by imposing standards and restrictions controlling its use, reflecting growing sophistication and awareness of the risk potential involved.

CASE 75. A woman was attacked while asleep but unable to describe the assailant. The prosecutor suggested hypnosis to facilitate recall. This was done by "a widely respected hypnotist" and as the victim relived the attack one of two detectives present "took control of the interrogation" and asked if the assailant was her ex-husband. She said it was but after hypnosis "she mistrusted her identification." The hypnotist and the detectives "encouraged her to accept her hypnotized account." The court expressed concern about police participation in hypnosis which "resulted in the witness passively confirming the state's hypothesis." Further, she became convinced it actually was her "genuine recollection" (Applebaum, 1984, p. 658).

People v Shirley, 641 P2d 775 (Cal 1982)

In this case the California Supreme Court excluded the entire testimony of a witness who had been hypnotized whether or not details of the testimony were obtained by hypnosis. The court reasoned that if hypnosis alters recall, and empirical data in 1982 proved it could, then the witness' memory of prehypnosis details could have been permanently changed. This was a blow to the rising use of police and lay hypnosis in the courts and set off a lively debate among professionally credentialled hypnotists (Diamond, 1983; Evans, 1983; Fulgoni, 1983; Watkins, 1983).

Diamond argued that even when hypnosis is practiced with the safeguards imposed by the courts there is no way to ensure accuracy of recall. Hypnosis "allows the witness to guess at" what occurred without the hypnotist "realizing that he is just guessing." Some will be accurate, lucky guesses but others "may be confabulations, fantasies or actual memories that properly belong to other experiences." Diamond was concerned about heightened susceptibility to direct and indirect suggestions of the hypnotist "of the kind and content which the subject believes is expected by the hypnotist." Regardless of the caution and competence of the hypnotist "the context and purpose of investigative hypnosis is in itself a power suggestion." The "uncertain witness is transformed into a confident one who withstands cross-examination. At best this is a form of improper coaching. At worst it encourages a witness to unwittingly commit perjury" and "once a prospective witness has been hypnotized the value of the witness as a competent testifier has been destroyed" (1983, p. 3).

Evans expressed the opinion that "most leaders in the field of hypnosis are fearful of abuses associated with attempted facilitation of recall particularly if not conducted by specially trained psychiatrists and psychologists." He explained witness' confabulation as the result of a "double bind situation where not to produce material would be tantamount to questioning the veracity of previous hypnotic involvement and the willingness to cooperate." There is a "great deal of pressure to recall."

In the writer's opinion, Evans described the stress experienced by the witness as much like a small child caught in an act not understood as wrong. Forced to explain, it is not at all unusual to hear of imaginary friends or ghosts or "cookie monsters." Adults are more sophisticated and Evans characterized this as a "cognitive flexibility" which under pressure from the court, by the crime and situation itself, lead a witness to "produce, merge, mix and invent memories" so that in relative desperation "created memories may take on a temporary reality." We see what we want to see, what we are prepared to see, what we need to see. Evans concluded "until there is replicated data that unequivocally shows that hypnotic memory can be reliably separated from accompanying embellishments" hypnotists should "proceed with a great deal of caution . . . there are many traps for the unwary" (1983, pp. 2, 6).

Fulgoni questioned the Shirley case as not accurately reflecting limitations of hypnosis and therefore not generalizable to all cases in all situations. He agreed that some witness testimony can be "suspect" by the effect of "vivid imagination" and "speculative rather than judgmental considerations" but the Shirley decision "is an overreaction to exaggerated peril." He contended hypnosis is still a viable alternative when police "have no knowledge regarding the details to be elicited" and "the elicited details are corroborated by subsequent evidence." This, Fulgoni suggested, would be a "double verification." Fulgoni concluded that "hypnosis should be used not as a matter of convenience but as a matter of necessity. Only when there is urgent need of immediate action or when the investigation has been stifled in all other respects should the police risk tainting a witness by employing hypnosis" (1983, pp. 5-6).

CASE 76. A rape victim briefly saw the attacker's car license number but could remember only the first three letters. Under hypnosis she remembered all the letters and numbers and even corrected herself—she had seen a C as a G. The defendant was apprehended and positively identified in a lineup by the victim. He gave conflicting statements but which contained some of the victim's observations and details. Fulgoni, who reported this case, points out that "under Shirley, the defendant would go free" and "without hypnosis he would undoubtedly never even have been caught" (Fulgoni, 1983, p. 5).

Watkins acknowledged that in the 1970s "hypnosis was heralded as the answer to a lawyer's dream" but that "experimental studies (Loftus,

1979; Hilgard & Loftus, 1979) report that memories are not recorded in the brain like a tape recorder," interrogators can externally and internally influence recall and fantasies, and "defendants can simulate and prevaricate under hypnosis . . . just as they often do in the conscious state." He observed: "In other words, there is both more wheat and more straw. Separating the wheat from the straw, the true memories from the contaminated ones is the crux of the situation."

Watkins concluded that while forensic hypnosis may have "often been badly practiced" this is "not a reason for banning it" any more than banning hypnosis because "there are bad hypnotherapists. The solution is to clean up our act. What has been shown is the need for competence and caution" and "when practiced in this way its rate of error may be no greater than is typical in the nonhypnotic interrogation of witnesses by police investigators" (1983, p. 7).

In letters to the editor of the APA Divison 30 Newsletter (August, 1983, p. 6) Joseph Barber and David Spiegel commented on the Shirley decision:

Barber concluded that "since it is possible for a witness to offer confabulated testimony, hypnotized or not, it is the responsibility of those who argue that we take hypnotically-elicited testimony as veridical to provide a means whereby the veracity of such testimony can be tested. In the absence of such means, what reasonable alternative can there be but to view hypnotically-elicited testimony with at least as much suspicion as any other testimony?"

Spiegel reported a case "in which a rape victim who had undergone hypnosis in the course of psychotherapy shortly after the assault was forbidden to testify against her alleged assailant three years later. In consequence, charges against him were dismissed. He is once again on trial in another jurisdiction for yet another rape. Thus, what has happened is that even therapeutic uses of hypnosis have led to the denial of a victim's right to testify against an assailant. The consequences of the Shirley decision are that even when hypnosis is used with the greatest of caution following every recognized guideline, the witness as a result is deprived of the right to testify, not just about his hypnotic recollection but about any recollection whatsoever of the events under consideration."

People v Hughes, 466 NYS 2d 255 (NY 1983)

In State v Hurd (1981) the court described errors in recall and unwarranted certainty as problems with victim or witness testimony with and without hypnosis. The New York Court of Appeals in People v Hughes excluded all testimony obtained by hypnosis, whether the facts were new or unique or given for the first time. It not only ignored but rejected the guidelines in People v Hurd. If other states follow this precedent it will in effect close the door completely on forensic hypnosis. Since the Shirley decision in 1982 continuing research studies have confirmed the possible unrealiability of hypnotically enhanced recall:

Dywan and Bowers (1983) presented 60 picture slides to 54 subjects divided into high and low hypnotic susceptibility groups) then tested recall three times with and without hypnosis. They found that "any pressure to enhance recall beyond the initial attempt may increase the number of items recalled but also the number of errors as well" (p. 184). They discovered that highly hypnotizable subjects recalled two times more than unhypnotized controls but with three times more errors. There was a "criterion shift" in the hypnotized subjects who exercised more subjective judgment, less caution, and a characteristic "vividness with which the subject is able to envision . . . possible memories" which leads to "a false sense of recognition" and "surprising certainty" of the hypnotically recalled material (p. 185).

Laurence and Perry (1983) screened 280 subjects with the Harvard Group Scale or Hypnotic Susceptibility, Form A, then screened those found highly hypnotizable with the Stanford Hypnotic Susceptibility Scale, Form C, yielding 27 very good hypnotic subjects, 16 females and 11 males, 21 to 48 years old. They created a pseudomemory (false) of being awakened by loud noise a week before the experiment. Thirteen of the subjects believed the fictitious event. Laurence and Perry concluded that victim or witness memory can be "modified unsuspectingly" and "an initially unsure witness or victim can become highly credible in court after hypnosis enhancement." This recall could result in "false but positive identification and to all of the legal procedures and penalties that this implies. Accordingly, the utmost caution should be exercised whenever hypnosis is used as an investigative tool" (p. 524).

In 1985, the **Council of Scientific Affairs** of the American Medical Association released a study by a panel of hypnosis experts convened "to evaluate the scientific evidence concerning the effect of hypnosis on memory." They concluded that hypnotically enhanced memory is not substantially different from nonhypnotic recall—it can be inaccurate, and is frequently a blend of correct and incorrect details. "Hypnosis can increase inaccurate responses to leading questions," and, as research has demonstrated, more errors are believed to be accurate by those who have been hypnotized. Further, "neither the hypnotist nor the subject can distinguish between actual memories and pseudomemories without subsequent independent verification" (p. 1921). The Council made several observations about hypotically enhanced recall:

> Hypnotic age regression is the subjective reliving of earlier experiences as though they were real" and "does not necessarily replicate earlier events" (p. 1919);
>
> The analogy of memory as being like a camera or a videotape recorder "is not consistent with research findings or with current theories of memory" (p. 1920);
>
> Hypnosis leads to "increased vulnerability to subtle cues and implicit suggestions that may distort recollection in specific ways, depending on what is connected to the subject" (p. 1922);
>
> The way a question is asked "can influence the response and even produce a response when there is actually no memory" (p. 1922);
>
> Hypnosis can "transform the subject's prior beliefs into thoughts or fantasies" accepted as accurate recall (p. 1922);
>
> Defendants with no special knowledge of hypnosis can stimulate or fake hypnosis or "wilfully lie" and "can deceive even experienced hypnotists" (p. 1921).

The Council recommends hypnosis be limited to the investigative process where "even a single correct recall may lead to important new evidence" provided incorrect responses would not interfere with the case or the use of hypnosis disqualify all evidence from the witness. Finally, it is recommended that only psychiatrist or psychologists conduct forensic hypnosis who "are skilled in the clinical and investigative use of hypnosis and aware of the legal implications in the jurisdiction" in which they practice (p. 1923).

FBI Guidelines

Ault (1979) described FBI guidelines for the use of hypnosis, based on a written policy adopted in 1968 and revised as needed. Hypnosis is used "as an investigative aid" when "witnesses or victims are willing to undergo such an interview." Only a "hypnosis expert" is to be used (psychiatrist, psychologist, physician, or dentist "qualified as a hypnotist"). A "specially trained agent (hypnosis coordinator) will participate in the hypnotic session." Sessions are taped in their entirety, videotape preferred. Informed consent is required prior to hypnosis at which time there must be "proper prehypnotic explanation of this technique to the witness or victim" (pp. 449-450).

Agents are cautioned that "information obtained through hypnosis cannot be assumed to be necessarily accurate. Careful investigation is needed to verify the accuracy of information obtained during these sessions." Only the hypnotist and the specially trained agent are to be present and the hypnotist supervises the hypnotic session and "must remain physically present throughout the proceedings." The FBI agent present "is qualified to question the witness or victim while under hypnosis but will not conduct the hypnotic induction or terminate the hypnotic state." It is likely the FBI has again revised those guidelines to keep pace with recent court actions."

COMPREHENSIVE FORENSIC HYPNOSIS PROTOCOL

From court cases which specify conditions and requirements for the use of forensic hypnosis, FBI guidelines used for over twenty years, those of professional hypnosis associations, and the writer's clinical and forensic experience, the following guidelines are offered.

Prescreening

Evalute the need and proceed only if hypnosis is the last resort, never if other alternatives would suffice. If hypnosis-enhanced testimony would stand alone without external corroboration it is not likely to be admissible and that witness' whole testimony may be stricken. If hypnosis-enhanced testimony would contradict or discredit other testimony of the same witness or another on the same side why attempt it? Do a cost-benefit analysis.

For optimal effectiveness, hypnosis should be used as soon as possible but never soon after trauma, severe stress or police interrogation. This is a "judgment call" by the hypnotist who may be caught between an urgent need for information and a subject's need of relief from suffering. To prevent hypnotist bias there should be no phone, mail or personal contact between the hypnotist or subject or others directly involved in the case (i.e., defendant or plaintiff, police, attorney, judge, jury, reporters, family or friends of defendant or plaintiff) except:

   (a) a letter stating the need for hypnosis and requesting the hypnotist decline or accept with a proposal specifying place, time available, and fee (usually per hour plus videotaping costs).

(b) any subsequent phone contacts to set or change dates should be logged as to persons speaking, date, time and length of conversation, or audiotaped and identified with date, time and persons speaking.

Preparation

If the hypnotist offers services by letter stating fees, place and tentative dates and times, the following documentation should also be included:

(a) Resume or vita. Ideal qualifications would be similar to those for any expert witness (current licensure of psychiatrist or doctoral-level psychologist perferably with a diplomate of five years hypnosis experience).

(Social workers, master's degree psychologists and registered nurses may have considerable experience with hypnosis, as do many police officers or stage entertainers, but they do not have comparable hours of formal education or of supervised clinical internships before hypnosis training and experience as do physicians and Ph.D. clinical psychologists. Only the most qualified, trained and experienced should conduct forensic hypnosis).

(b) The written proposal should include services to be provided, a confirmation of no conflict of interest or prior personal or professional contact with defendant, plaintiff, witnesses or friends or family of them, police officers, judge, or attorneys involved in the case, the place, dates and times available, and stipulations as to whom will be present (only hypnotist and subject; others documented with rationale and extent of participation), provisions for videotaping all sessions from the moment hypnotist and subject meet to the moment the subject leaves the room, to whom the videotape will be given with appropriate signed release of information, backup copy to be retained for a specified time by the hypnotist, required written transcript or audio or videotape of the subject's previous statements relevant to hypnosis recall.

(c) The subject should have a mental status examination by someone other than the hypnotist, the report in writing forwarded to the hypnotist prior to the hypnosis session, or, if time does not permit, communicated verbally to the hypnotist with date and time logged or the report audiotaped and identified by date.

(This is to assure legal competence and to inform the hypnotist of the subject's mental status by a third party to avoid charges of bias or conflict of interest).

(d) The subject should receive a copy of the hypnotist's proposal as part of informed consent. It is helpful, too, to provide a photograph of the hypnotist since they will not meet until the actual hypnosis session. A booklet, pamphlet, article, or reprint of information such as these guidelines should be given the subject at this time. The referring party, family member, or significant other to the subject should clearly state that:

Hypnosis may or may not be successful;

It will be conducted by a **qualified, licensed, experienced professional;**

It **will not be a danger** to the subject's mental health; any adverse side effects will be treated by the hypnotist or referred to an appropriate, qualified professional;

There will be time beforehand to ask questions, prepare, and feel comfortable;

The subject is **free at any time to stop** the hypnosis without penalty (i.e., not used by either side to discredit other testimony, agreed in writing by both sides) and the subject will be instructed how to signal to stop hypnosis (verbal and ideomotor signalling).

The Hypnosis Session

The place should be a comfortable, safe, neutral environment, not a police station or attorney's office, preferably the hypnotist's usual treatment setting.

No one should be present except the hypnotist and the subject except under documented special circumstances (i.e., parent or guardian for a child, mentally retarded or disabled person who cannot communicate well, artist if a sketch is required, a technical expert, etc.) and any departure from this rule must be approved in writing by both sides.

Ample time should be provided to complete the hypnotic interview, allowing for briefing, questions, confirmed informed consent, initial conscious recall, hypnotized recall, dehypnosis, debriefing and termination or scheduling the next session.

Documentation should be maintained, confirming the proposal, schedule, fees, informed consent, agreement or contract, receipt of and review of background information and the subject's prior statements, and a record of the date and starting and ending time of each session. If more than one session is required there should be no personal or professional contact between hypnotist or subject between sessions.

Technique: There will be no leading or open questions or misleading statements (i.e., that the mind is like a camera or tape recorder and can remember everything) and questions should be so worded so as to encourage as much as possible the subject's own ongoing narrative account, with minimal interruptions or questions and these should be solely to facilitate detailed recall, devoid of interpretive statements or value judgments.

Prehypnosis recall. Before hypnosis is begun, the subject will give an uninterrupted account of what is remembered about the event or incident in question.

Hypnotic recall. During hypnosis, the hypnotist will ask simple, direct questions, refrain from leading or open questions, opinions or value judgments or changes in voice tone, emphasis by word choice, repetition, pause or rate of speech, approving or disapproving words or statements, or any verbal or nonverbal interventions other than to objectively facilitate recall. If, for documented reasons, it is necessary that anyone else is present and needs additional information, questions will be written and given to the hypnotist who is the only person who will ask questions.

Posthypnotic recall. After hypnosis, but before the end of each session, there will be posthypnotic recall and differences between recall before and during hypnosis will be pointed out to clarify the scope and degree of additional detail to prehypnotic memory.

ANTISOCIAL ACTS

The possibility of a hypnotized person committing antisocial or illegal acts during or after hypnosis has been a controversial subject, debated inconclusively by clinicians and researchers. The consensus is that it is highly unlikely but a person could be deceived or misled into antisocial acting out. Such an individual would be highly suggestible (not necessarily a good hypnotic subject) or believe or be led to believe that hypnotic suggestion is irresistible and overpowering. Bramwell wrote in 1903: "If a subject believed that hypnosis was a condition of helpless automatism, and that the operator could make him do whatever he liked, harm might result, not through the operator's power, but in consequence of the subject's self-suggestions" (p. 425). Bramwell concluded that "hypnotism, through ignorance or malice on the part of the operator, might be misused as to do harm" (p. 427).

In reviewing research on this subject, Barber (1961) observed that "when a literate individual in Western culture is told that he is to be 'hypnotized' he is by implication being informed that he is not only permitted but also expected to carry out performance which he would otherwise inhibit." Even using the word hypnosis "may also contain a concomitant message" that when the subject "abdicate(s) control of his behavior, the hypnotist is responsible for the consequences" (p. 111). Orne (1962, 1972) described the complete cooperation and trust of hypnotized persons in researchers, and Orne felt they do so in the belief they are helping in the advancement of science.

For whatever reason, hypnotized subjects in experiments have stolen money, lied, read mail of others, thrown what they thought was acid and pointed what they thought were loaded guns at others, and reached into a box thinking there was a snake in it (Barber, 1961; Orne, 1972). In comparison, few unhypnotized controls or hypnosis simulators did so. The following unfortunate cases involved behaviors not anticipated or far in excess of what was expected:

CASE 77 was an army private who participated in hypnosis experiments before. Placed in trance he was told that when he opened his eyes he would see "a dirty Jap soldier. He has a bayonet and is going to kill you unless you kill him first." A lieutenant colonel stood directly in front of him when he opened his eyes. It took three men "to break the soldier's grip, pull him off the officer, and hold him until the experimenter could quiet him back into a sleep condition" (Barber, 1961, p. 307).

CASE 78 was a young man who stole a gun and used it in an armed robbery. He had no prior criminal record, was in good physical health, never before referred or treated for mental problems. Several hours before stealing the gun he was hypnotized by a stage entertainer and told he was "the best cowboy in the Wild West," a quick draw expert and a crack shot. He remembered returning to the audience after being hypnotized "confused, restless and as though there was something missing in his head." Mental health professionals treating him afterward concluded his bizarre behavior was due to "some internal, previously unfulfilled psychologic need" satisfied when "previously adequate situational controls were removed" by hypnosis (Kleinhauz & Beran, 1984, p. 286).

CASE 79 was a man who attempted to choke his wife while she was sleeping. "A few days previously he had been successfully hypnotized for the relief of bruxism." His "underlying anger" (at his wife) "was now no longer held in check. As an emergency measure, he was rehypnotized and his bruxism was restored" (Rosen, 1962, p. 671).

In commenting on the above case, Rosen advised that "all regression should be controlled. The patient otherwise may conceivably be harmed. If treatment is symptomatic, then, before symptoms are suggested away, some of the functions they serve should be determined" (Rosen, 1962, p. 671).

Some individuals dream or daydream of being someone else or being somewhere else in the present or at some other time in history. Movies and fiction books and magazines help satisfy these largely unconscious inner needs and wishes. For this young man, hypnosis provided him a brief interval when his fantasy, dream or wish became a reality. Barber (1961) studied reports of antisocial acting out during or after hypnosis published between 1937 and 1958, and he concluded: "If 'hypnosis' played a role in these cases" it was "in reinforcing and extending a delusional system which had preexisted, and in providing the subject with a rationale for justifying his behavior to himself and to others" (p. 119).

It is not known how many prehypnotic subjects are sufficiently obsessive in convictions or interests to be susceptible to antisocial hypnotic suggestions. How can hypnotists identify high risk subjects? Practical suggestions for risk management would be to carefully observe subjects before and during trance, thoroughly debrief after hypnosis, and to avoid any suggestions, symbolism, or hallucinated situations which could directly or indirectly be associated with antisocial acts. If there is the remotest possibility of misperception or antisocial acting out, subjects should be told before, in and after trance that they are responsible for their actions after hypnosis just as before. The best preventive practice would be to include this instruction routinely in every induction.

SUMMARY

Clinical, experimental and stage hypnotists can unknowingly contribute to or precipitate unwanted side effects by personal or professional risk factors. Personal factors are traits, attitudes and verbal and nonverbal behaviors unique to their personalities. Professional risk factors are a lack of knowledge or skill, poor screening, history or observation, missed cues or misdiagnosis, ambiguous of confusing suggestions, poor timing, inappropriate choice of words or imagery, poor judgment, ineffective dehypnosis or debriefing, and contributing to antisocial acts.

Replicated experiments over many years demonstrate how subtle cues can shape subject responses and posthypnotic behaviors (demand characteristics or E-bias). More recent research data on hypnotically enhanced recall demonstrate how memory can be distorted or irrevocably contaminated (forensic hypnosis). Court decisions are restricting the use of forensic hypnosis. Detailed guidelines should be followed to ensure legal competence (mental state examination), informed consent, briefing to dispel myths and misconception, a free-running narrative recall with unobtrusive support (no leading questions) followed by a simply structured recall to elicit details (simple questions only), debriefing or posthypnotic discussion (must compare prehypnotic, hypnotic, and posthypnotic recall and differentiate them), and re-entry to ensure the subject's continuing mental health and stability (arrange for any follow-up care) and to do all these in such a manner so as not to jeopardize the side on which hypnotically-enhanced testimony is used.

CONCLUSIONS

It is recommended hypnotists ensure they have an adequate knowledge base (psychodiagnotics; symptomatology of mental disorders), sharpen skills (observation; crisis intervention), avoid the pitfalls and problems cited here, and integrate preventive methods routinely in their use of hypnosis.

CHAPTER 4

ENVIRONMENTAL RISK FACTORS

Where it happens
can be as striking
as what, why and how

— Anonymous

WHERE HYPNOSIS occurs is a significant variable and can be a risk factor if elements of the physical setting are perceived negatively by the subject. Environmental variables can influence and affect hypnotic induction in clinical, research and stage entertainment settings. While not complications themselves external cues such as background sound, uncomfortable furniture, even colors or pictures on the wall or what the hypnotist wears can distract rather than enhance concentration, add to resistance and render the subject less accessible to the hypnotic process. Interruptions such as ringing phones or doorbells, traffic sounds, overheard conversations, paging systems, or even weather (wind, rain, thunder) can be unknowingly integrated into a subject's hypnosis experience.

Physical discomfort can interfere with the hypnotic process. Common sources are crossed legs, folded arms, clasped hands, tight or itchy clothing, eyeglasses, chewing gum or candy, an itch not scratched, the need to go to the bathroom or an unusually hot, stuffy, cold or drafty room. It is difficult to relax at a hypnotically suggested ocean beach on a clear, sunny day with ringing phones, people talking, thunder or trucks backfiring nearby. "Whatever's used can be abused" is the aphorism for potentially negative environmental risk factors.

Mesmer was keenly aware of the importance of the physical environment to hypnotic induction. He routinely used soft background music, diffused light, an impressive costume of robes and regalia, an imposing, commanding presence, and the famous baquet of iron rods in the wooden tub of water. He experimented with environmental factors, with colored light—he found blue light more effective to trance induction than red. Stage hypnotists are more attentive to these "props" than clinicians or researchers. In 1876, Ralph Waldo Emerson commented on the importance of physical appearance and overt behaviors: "What you are—thunders so that I can't hear a word you say."

One of the reasons researchers have difficulty replicating experiments is inadequate reportage of physical factors of the research setting. The room where an experiment is conducted can vary significantly in its size, furniture, furnishings, acoustics, temperature and humidity. Few research reports include such information. Tight clothing, body posture, time and timing can lead to sufficient discomfort to affect attention, concentration and motivation. The "pins and needles" of temporary circulatory problems is not only uncomfortable and distracting but for some subjects in trance can become somatic delusions or misperceived as the hypnotist's touching or restraining them.

There are environmental risk factors completely independent of the treatment and research settings. "No one is an island," as John Donne observed in the 14th century. No one lives entirely alone or "in a vacuum." There is for each of us an inner world of subliminal fantasies and myths, of consciously obtained not fully tested or understood misinformation and misconception. This residue of "subjective truth" is the product of what is and has been said around us, claims and counterclaims in advertising and promotion, radio, TV and movies, books and articles, the thoughts, feelings and opinions of neighbors, friends, family and coworkers. These influence subjects' expectations of hypnosis and help create demand characteristics.

Existential (here and now) environmental risk factors can be classified according to the sensory system involved.

VISUAL STIMULI

The visual environment has psychological impact, in terms of the hypnotist and the room in which hypnosis occurs. What hypnotists wear projects much of their personality. Dress too casually and the subject may suspect a casual attitude toward services to be provided. Dress too conservatively and the subject might interpret this as distancing, an elitist, condescending attitude. Hypnotists should be aware of the "image" they may project with what they wear. The author once had need to refer a guilt-ridden woman to a priest because she had great need to be forgiven by God or someone who symbolized God. She had not been to church for ten years. An appointment was made. The woman stood on the rectory porch anticipating the opening door to disclose the living symbol of the Almighty. The priest opened the door smiling broadly wearing slacks and what the woman described as a "garish Hawaiian shirt." She had to go elsewhere to assuage her guilt!

Waiting rooms, treatment rooms and labs also have visual impact. The atmosphere can be magical and mysterious, businesslike and impersonal, comfortable and comforting, busy or boring. The author has visited a great variety of clinical and research settings. Much can be done to improve the "feel" of them. One colleague has a large wall poster of Che Guevara! Another has an office full of museum-like artifacts. Still another has no less than a dozen house plants, some growing wildly, the room resembling an Amazon rain forest. Some floors are tiled and "hard." Some furniture looks uncomfortable or forbidding.

Lighting is an important environmental variable with psychological impact. In novels and movies of military or police interrogation, the prisoner squints into a bright bare light bulb. The patrolling policeman flashes a light into dark doorways. Indirect lights in a waiting room are more "welcoming" and restful than direct ceiling lights. Every treatment or research room should have a dimmer switch convenient to the hypnotist so that room brightness can be diffused during hypnosis. A practical test is to sit in the subject's chair with eyes closed. There should be no light evident through the eyelids, no difference with eyes closed with lights dimmed or lights out completely.

CASE 80 was a male college student who volunteered in a hypnosis experiment. He failed to awaken from the hypnotic dream item on the *Stanford Hypnotic Susceptibility Scale, Form C.* He had no similar difficulty with other items on either the Stanford or Harvard scales. He complained "hypnosis would have been more effective if all lights in peripheral vision had been turned off." The lighting was the same for all subjects for that item and "his disturbance by the light appeared to have special significance." He described his hypnotic dream as like a horror movie he saw at age 5 or 6 in which "a typical monster type" hypnotist wearing a black cape used a flashing light "to hypnotize people and they never came out of it." (J. Hilgard, 1974, p. 290).

Josephine Hilgard, in reviewing this young man's difficulty emerging from trance, observed that there was "a strong possibility the dream in some manner defined hypnosis for him, that he became more deeply hypnotized by the flashing light in the dream . . . the threat of the movie hypnotist that he could not come out of hypnosis became actualized for him. If this conjecture is correct he was hypnotized during his dream with a spontaneous deepening corresponding to the deliberate use of the dream as a deepening device" (1974, p. 291). The "dream within a dream" deepening method is used routinely by the author in a structured standard of treatment called "the envelope technique" described in the next chapter.

CASE 81 was a female subject recently divorced and treated with hypnosis as an adjunct to individual psychotherapy. Hypnosis was used routinely during most of the one-hour sessions. Toward the end of her fifth session, the hypnotist complimented her for "opening up and letting her real feelings flow." Bewildered, she thanked him but said she didn't feel the session differed from those of previous weeks. "Ah," the hypnotherapist said, handing her his hankerchief, "dry your eyes. Crying is different, isn't it?" She blushed in mild embarrassment and replied: "Oh! That's from my contacts. I just started wearing them this week." The hypnotist was very observant but not too perceptive (Author).

AUDITORY STIMULI

There are two aspects of auditory stimuli important to hypnosis: sound and noise. Sound is that aspect of auditory sensing that can facilitate induction when used judiciously. Noise is an interference and a distraction which should be reduced or eliminated.

Sound. Hypnotists should check the acoustics of every room. Draperies, carpeting, and padded cloth furniture absorb sound more than hard tile or wood floors, venetian blinds, or plastic or wood furniture. A "white sound" generator is helpful not only because it is restful but as background sound it masks and blocks noise from outside the room. Some sound generators have controls so that volume and tone can be adjusted to room size and subject hearing (auditory acuity varies individually) and there is a choice of ocean, rain or waterfall effect. Busy hypnotists may not note the setting of the machine or the type of sound used. Playing rain to a subject who has previously been hypnotized with the rise and fall of ocean surf is an environmental risk factor. Using the correct type of sound but blasting too loud or so low as to be inaudible changes sensory input and departs from standard laboratory and treatment conditions. It is very important to note the type of sound used and the volume and tone settings on the subject's file.

Noise. There are many sources of unwanted sound (noise). Radio, TV or voices from the adjoining room or elsewhere in the building, traffic sounds from the street, telephones or paging system, a knock on the door, even weather (wind, rain, thunder) are common noise sources. Noise can become a risk factor depending on the subject's associations. "Hearing voices" is a classical sign of psychopathology, and a subject in trance who actually hears voices from outside the room could become greatly upset and agitated if unable to connect them to reality. Hypnotists who forget to disconnect the phone in the room used for hypnosis are apt to jump a few inches off the chair along with the subject if the phone rings.

As was pointed out, it is difficult to relax on a peaceful hypnotic ocean beach, under the warm sun, in the fresh air, when there are noises like trucks, motorcycles, buses, trains, airplanes, horns and bad mufflers sounding nearby. Research subjects may be annoyed, then resistant, if instructions are read aloud to them in a boring, bureaucratic voice, not given an opportunity to ask questions, not listening to and with poor eye contact.

CASE 82 was a female subject treated with hypnosis for habit control, excessive cigarette smoking. She chose an ocean beach as the most relaxing visual imagery. During the first induction there was a loud thunderstorm, the sounds of which penetrated through the hypnotist's background white sound, even when the volume was adjusted upward. When she came out of trance she complained that she not only heard the thunder but there was a storm on the beach. Rehypnotized, during which time the actual storm outside the room subsided, she again experienced a storm on the beach. During her second session, with clear weather outside, she reported no storm on the beach "but it was overcast, gray and threatening and the wind was blowing so hard I could feel the sand sting my face." Rehypnotized, the weather moderated. In subsequent inductions the weather on the beach improved to bright sun and cloudless sky—but there was always a strong breeze (Author).

The auditory stimuli of the thunderstorm "imprinted" her sensation and perception and became an integral part of the hypnosis process. Incidentally, she managed to stop smoking and found autohypnosis relaxing. To fall asleep at night she would picture the beach, take a deep breath or two, and drift off to a restful, uninterrupted sleep. Several months after treatment was terminated she reported that she continued to use autohypnosis for relaxation and to fall asleep at night. Asked about the wind, she replied: "Oh yes, it's always breezy there but I've learned to really like it—the fresh air is good for me." It may be that at some deeper level of consciousness the storm symbolized the force needed to overcome her smoking habit and the strong breezes were "the winds of change."

TACTILE STIMULI

Tactile sensing includes responses to heat, cold, pressure and vibration. A warm room might facilitate the mental imagery of an ocean beach but a cool room could inhibit it. Tight, uncomfortable clothing can restrict blood flow or breathing. Body position can impede blood circulation (crossed legs or arms). Sitting in a plastic chair in a cool room is not conducive to physical or emotional warmth nor is sitting upright in a hard-backed chair.

The ideal tactile environment for hypnosis is more "soft" than "hard." Carpeting is better than a hard tile or wood floor. Soft, large, comfortable cloth chairs are better than plastic or leather. A recliner chair is very good. Pillows should be provided to allow for subjects of varying height, to provide a clear, open airway. An afghan placed over the subject from neck to toes makes some subjects more comfortable, physically and psychologically. As a general rule, subjects should not cross their legs, fold their arms or clasp their hands. Subjects should be encouraged to loosen any tight clothing, remove eyeglasses unless they are used to wearing them most of the time, and remove any jewelry that might be uncomfortable or interfere with breathing (i.e., some women wear what quite appropriately are called "chokers"!).

CASE 83 was a group of teenagers huddled on the beach at Atlantic City on a cool, windy fall afternoon. A hypnotist was strolling along the beach and asked the group why they were not swimming. "It's too cold," they said. He offered to hypnotize them "so that it'll be warm if you really want to have a swim." They agreed, he hypnotized them, and they ran laughing happily into the surf. The hypnotist's smile of approval turned to a look of serious concern when he saw that the youths continued to swim in a straight line away from the beach. Lifeguards down the beach blew whistles and yelled at the swimmers but they did not turn back. Much to the hypnotist's embarrassment, boats were required to retrieve the swimmers. The hypnotist overcame chilly weather but got a very cold reception from the beach patrol (Author).

CASE 84 was a male college student who volunteered to be hypnotized in a stage presentation at his college. A good subject, he was selected to go on stage and did so. He was seated in a steel folding chair just a few feet from the edge of the stage. The entertainer-hypnotist put the young man in trance and before he could be given any instructions, he "passed out," fell off the chair to the stage floor, and rolled off the stage, falling into the orchestra pit. This case is included to show how a hypnotist was unaware of the very real risk factor in the physical environment. It had nothing to do with the hypnotic induction, which was very successful, but if the young man had not been hypnotized he would not have fallen off the stage (Echterling and Emmerling, 1984).

SMELL AND TASTE

There are distinctive smells most people associate with specific places: a garage, cellar, library, church or synagogue, garden, the seashore. There are characteristic odors of foods and objects: tobacco smoke, cabbage, perfume or after shave lotion, onions or garlic, flowers, gasoline. Miller (1979) used smell in hypno-aversive habit control to help subjects stop smoking. He kept under his desk a jar containing about a glass of what was once water into which he put cigarette and cigar butts. The jar was several years old and it smelled awful. In his desk drawer he kept a small bottle of floral scented after shave lotion. When the subject was in trance with eyes closed he would unscrew the lid of the jar and as he told them "cigarettes will never taste the same, they taste awful . . . " he would waft the evil mixture under the subject's nose. "But when you don't smoke," he continued, replacing the stale tobacco "stew" with the floral scent, "the air will smell fresh, like a flower garden." It was an effective technique, using smell as part of posthypnotic suggestion.

CASE 85 was a middle-aged woman, a Mother Superior in a Roman Catholic religious order, treated with hypnosis for control of colitis. The hypnotist enjoyed a cigar occasionally after dinner. His office adjoined his home, in the same building. He had finished "a good smoke" seated in his office, then rose to greet his next patient. It was the nun, and he had forgotten she was the next appointment. He said nothing, proceeded with hypnosis, and she was able to attain a much deeper trance state than ever before. Afterward, during debriefing, they discussed her good response. She smiled and said: "It was the cigar." She explained that she felt very close to her father who enjoyed a cigar after supper. The aroma of the cigar proved a strong positive factor specific and unique to this particular subject (Author).

CASE 86 was "a mature woman" who was given suggestions of "well-being and happiness" while presented with the smell of perfume. She "wept profusely both in the trance and afterwards—the perfume had stimulated the recall of a bouquet of flowers at a funeral" (Williams, 1953, p. 6).

For several years, the author used a variety of scents to facilitate visual imagery in trance, including a jar of stale tobacco, ashes and water for hypno-aversive conditioning for habit control of smoking. There were a surprising number of persons who associate floral scent with funeral homes and the elicited sad feelings or memories interfered with treatment goals. The mind's "memory bank" has many "deposits"—happy and sad, and certain hypnotic techniques are like a roulette wheel which can unintentionally and randomly "hit" an unpleasant memory. In such cases, rehypnosis is the most effective way to neutralize, further vent, or treat the situation. Which of these is effected depends on the significance of the recalled event and its effect on the subject's mental state, needs and wishes.

Buddha considered the senses to be "the windows of the mind" and external cues from the physical environment pour through these "windows" into the mind. In addition to the five senses, there are other external factors which can interfere in hypnosis. A subject who is not feeling well may have difficulty following instructions. A person who has a bad cough, sneezing from a bad cold, breathing difficulties from asthma or allergies may attend more to keeping a clear airway than relaxing. There are subjects who experience dry mouth in hypnosis, who wet their lips or clear their throat frequently. It is helpful to have a glass of water nearby for them.

The author has heard many complaints from subjects hypnotized by others which the reader may find difficult to believe. Subjects have complained about family members of hypnotists walking into the treatment room during trance and conversing with the hypnotist about some trivial matter. Some hypnotists answer the phone while the subject in trance recalls the conversation later, not fully aware of its meaning. The subject might not only hear the telephone conversation but its content could easily contain trigger or stimulus words to precipitate complications. Hypnotists who answer phones in this manner may not continue exactly where they left off in the hypnotic process, may omit important information and instructions, or be other — occupied and no longer fully attentive to the task at hand.

Third party payors, peer review committees for clinicians or **department heads** or **research review committees** for researchers, can exert subtle or direct pressure on hypnotists who may then become therapists or researchers "in a hurry," limiting available time for careful induction, moving too quickly over sensitive areas, missing cues and clues, and investing more time in paperwork than the treatment or research process.

There are **interpersonal variables** involved in the physical setting such as failure to debrief external cues (e.g., background sound, words or phrases used to deepen trance or in trance which may be inimical to the subject's life or work situation). An accountant or someone operating equipment with digital readout who blocks a numeral because of incomplete dehypnosis is more accident and error prone. Driving or operating machinery can be hazardous if there has been incomplete dehypnosis to sounds used in hypnosis similar to those in the workplace.

Subjects incompletely dehypnotized describe "strange deja vu feelings . . . a spooky creepy feeling . . . spaced out . . . groggy . . . stoned . . . dizzy" and these feelings persist hours to days after hypnosis. Experiencing these feelings may not be as dramatic as falling off a stage or hysterical weeping when smelling perfume but it is partial hypnosis. While it is a mild complication the fact that it is unnecessary and potentially dangerous should be of concern to all hypnotists. Following "standard procedure" from checklist format in a structured risk management system would greatly reduce such after effects by ensuring dehypnosis, rehypnosis, debriefing, re-entry and followup.

What follows is bias, the author's perception of the **nature of hypnosis,** based only on his own clinical and research experience. The setting in which hypnosis occurs is very important. It can have an influence on the severity of complications depending on what the environment and external stimuli mean to the subject at the time. Clinical, research and entertainment settings each carry its own type and degree of risk which vary also with the hypnotist, the subject and even within the same subject from time to time. Complications risk is proportionate to the impact of and the involvement in the hypnosis experience. That risk is relative. Individual hypnosis can have heavy impact because of its 1-to-1 personal involvement. Group hypnosis is less personal but the risk lies in the impossibility of observing every subject every minute. Stage hypnosis can have the heaviest impact because a vulnerable subject is stripped mentally naked publicly to the world, the audience, society in miniature. It is a reversal of all the positives of group psychotherapy, without confidentiality, informed consent, screening, thorough dehypnosis, debriefing or followup.

Physical, Emotional and Neuropsychological Factors

Comfort. A safe, comfortable environment symbolizes stability and security.

Comforting. A friendly, accepting hypnotist symbolizes caring and approval.

Multisensory involvement facilitates a more complete, holistic involvement physically and mentally. Hypnosis involves a majority or unanimous vote of the senses, as many "open windows" to the mind as possible.

Regression. Physical relaxation (comfort, nonverbal) begins a process of mind (brain) and body regression. The mind (cerebral cortex) relaxes its grip on reality testing and critical thinking. The body (autonomic nervous system) continues to relax, becoming more vagal (parasympathetic), the reason early theorists called hypnosis "sleep" or "sleep like" and also "suggestion" or "seduction." The extensive clinical literature on the use of hypnosis in psychosomatic disorders establishes autonomic involvement.

Subcortical. As regression continues, cortical function is less predominant and subcortical involvement increases, from hypothalamus to the limbic system, deep in the "ancient brain."

Hypnosis is a complex mental process, both cortical and subcortical, conscious and unconscious. Previous theorists defining hypnosis as suggestion, regression, a unique subjective or altered state are all correct. It is all these and more, the ancient brain sharing with the modern brain, but it is more subcortical than cortical. The use of the word "sleep" and "relax" suggest a vagal autonomic function and dreamlike hypnotic hallucinations point to the pons, the "dream center" in the brain stem. Kleinhauz and Beran (1981) described the dramatic case of a young woman rendered mute and catatonic after being hypnotized (Chapter 1, page 1). The symptoms suggest that hypnosis unleashed or hypnosis is a powerful and also very primitive force that regressed the individual completely to a preverbal level of existence, totally insulating her from reality. it was for her the ultimate defense, no defense whatever, limbic limbo. Trance logic has characteristics of this distancing from "normal" reality testing and from critical thinking and judgment but it is still more "civilized" than the case described by Kleinhauz and Beran.

SUMMARY

Clinical and research hypnotists each strive for standards of practice specific to the setting. Clinicians seek to establish and maintain "standards of care" and these are best achieved in a structured or "standard treatment setting" with routine procedures. Researchers try to provide "standard laboratory conditions" to prevent contaminating and extraneous variables so that the data flows freely and naturally to prove or disprove the hypothesis. Both settings should provide a physical and also a mental environment which are safe and free of risk.

Furnishings should be pleasant or at least neutral psychologically, friendly not foreboding. Furniture should be comfortable. The room should have good acoustics with minimal distractions. Temperature and humidity should be in "the comfort zone" for the climate. Sensory input should be carefully considered and optimally used (sound, smell, light). Hypnotists should design the physical environment like a theatrical production, attentive to choice of and the placement of furniture and furnishings, sound levels and noise suppression.

CONCLUSIONS

As behavioral research clearly demonstrates, the environment is a major factor in personality development and in shaping social-interpersonal behaviors. We are social beings. Wherever hypnosis is used, objects, equipment and materials in the physical environment can positively or negatively influence the hypnotic process. Effective hypnotic induction is a multisensory experience in which there is "something for each sense" and also something for the mind and for the body.

Hypnosis is a gradual progression conscious to unconscious, from predominantly cortical to increasingly subcortical brain function.

Hypnotists should be aware of the involvement of sensation and perception in this process. The physical environment should be comfortable and the emotional environment should be comforting. Anything in the physical environment can become a negative influence if the subject has painful or unpleasant associations to it. For this reason there should be a clinical consultant in every hypnosis research project, readily available to subjects in need or preferably who debriefs every research participant. When subject risk factors are present and combine with hypnotist risk factors and when both of these occur in an uncomfortable physical setting, the likelihood of complications is greatly increased. "Whatever's used can be abused."

CHAPTER 5

RISK MANAGEMENT

*Anything that can go wrong
will go wrong!*

—Murphy's Law

*1st corollary:
Murphy was an optimist!!*

SURELY, no one intended the meltdown of the Russian nuclear reactor at Chernobyl in April of 1986. The same is true for the Challenger spacecraft in January of the same year. These two "accidents" occurred in scientific, very sophisticated settings: the nuclear power industry and space exploration. It was Murphy's Law at work again and this same potential for danger exists for hypnosis, in clinical, research or stage entertainment settings, with varying degrees of risk specific to the setting, subject and hypnotist.

It is also unfortunately true that accidents can be prevented, by further inspection, quality control, preventive practices and effective risk management. This is true for the nuclear power industry, space exploration and also for hypnosis. It is possible that a severe complication might result regardless of the circumstances, that the complication would result anyway. Unless and until it can be proven that complications are inevitable, irreversible and untreatable, hypnotists should make every effort to routinely use preventive practices and risk management. If complications are avoidable, hypnotists need to ensure that they are aware of and use preventive practices routinely.

What makes hypnosis risk management difficult is the multicausational aspects of behavior and the complexity of human personality. Very little skill is required to hypnotize—even children have done so without any training whatever. Real expertise is needed to be aware of problems as they develop, to intervene promptly, neutralize their effect, and provide further treatment if required. These are areas in which lay hypnotists are lacking, in view of limitations in education, training and supervised experience as compared to licensed physicians, dentists and psychologists. Lay hypnosis organizations and associations consider this observation to be "empire building" or a "turf issue" where the professions attempt to "corner the market." Most professionals would lose very little income if they discontinued hypnosis completely from their practices. Their concern is an ethical one, in the words of Hippocrates 2500 years ago, "to do no harm to the patient."

WHO SHOULD HYPNOTIZE?

For a hundred years hypnosis experts have warned about the potential dangers or hypnosis misuse. In 1887, Frederick Bjornstrom cautioned that "there lies such an infernal power in the hands of the hypnotizer that every one ought to be strictly forbidden to meddle with hypnotism except those who assume the responsibilities of a physician and who have the people's welfare and woe in their hands" (p. 415).

Rosen (1960) wrote "the self-styled hypnoanalyst without years of psychiatric (and analytic) training can be a dangerous person . . . and so can the self-appointed teacher of the subject. No one, in fact, should ever treat patients on hypnotic levels with techniques beyond the range of his usual professional competence with unhypnotized patients" (p. 686).

In 1962, the Group for the Advancement of Psychiatry, in its position paper held that hypnosis is appropriately used in therapy "only when its employment serves therapeutic goals without posing undue risks to the patient." Sedative, analgesic and anesthetic purposes, anxiety relief and symptom suppression are approved uses of hypnosis but "on a still more highly selective basis, as an adjunct in the treatment of patients with neurotic or psychotic illness" (p. 704). This recommendation, and that of Rosen just quoted, point up the need for hypnotists to practice only in the area of their competence. Medical doctors, with their many years of education, training and supervised experience, are reminded that treating neurotic or psychotic patients is "more highly selective," implying the need for further specialized training.

Miller (1979) warned that "the patient in trance is in a highly vulnerable state and can be subject to considerable anxiety, guilt and emotional disturbance. Great care must be exercised not to unduly disturb patients in this state and to avoid creating conflicts and symptoms which may later be difficult to eradicate." He concluded "no one should be permitted to use hypnosis to treat emotional illness unless he has been adequately and properly trained in psychodynamics and has clinical training under competent professional supervision. Hypnosis is a potent instrument which cannot only relieve symptoms but create them" (p. 331).

Gravitz et al. (1982) recommended that "while clear consensus on risks and dangers of hypnosis is not possible, there should be general agreement that the hypnotist must meet the basic ethical and professional requirements of appropriate competence and integrity" (p. 299). Echterling and Emmerling (1984), after reporting their research findings on casualties from stage hypnosis, concluded that "hypnosis is a powerful technique that can dramatically affect our consciousness, experiences and behavior—for better or for worse. It should be practiced only under the most controlled circumstances, by ethical and qualified practitioners, and with individuals who have given freely their informed consent. Without these minimal conditions, sponsoring institutions may be held liable for the physical injury or psychological trauma that some members of the audience are likely to experience" (p. 14).

There are dissenting opinions. Kroger (1977) wrote that "no one ever died of hypnosis." The case described in the opening pages of this book came very close to death (Kleinhauz and Beran, 1981). Elsewhere in his book, Kroger acknowledged possible problems which might arise from "prepsychotic personalities," organic conditions "more amenable to medical therapy," symptom substitution and complications arising from a subject's "wishful thinking, deceit or actual fabrication." Most textbooks and most hypnosis courses and workshops minimize the dangers of hypnosis, possibly because few physicians, dentists or psychologists encounter severe complications. They are fortunate—but not immune.

Who then should hypnotize? This question is better worded: Who is best qualified to hypnotize with least risk? Risk management and preventive practice involve as much judgment as skill, as much awareness and perception as ability and practice. In any professional or craft, art or science, these traits emerge from a blending of formal education (the knowledge base), training and supervised experience (skills base) and professional licensing (demonstrated competence by written, oral and practice examinations), binding the practitioner to strict legal and ethical constraints. Physicians, dentists, psychiatrists and psychologists have met these requirements before they study hypnosis, uniquely preparing them for specialization in hypnosis.

Licensed physicians, dentists, psychiatrists and psychologists who choose to add hypnosis to their therapeutic repertoire attend postgraduate courses or workshops to acquire knowledge, skill and ability in the theory and practice of hypnosis as it relates to their profession. After five years of experience, they can apply for written, oral and practice examination for a Diplomate from their professional Board (i.e., American Board of Dental Hypnosis; American Board of Medical Hypnosis; American Board of Psychological Hypnosis).

MQH (**Most Qualified Hypnotist**) is a stereotype, one who is a licensed medical doctor, dentist, psychiatrist or clinical psychologist with a Diplomate in hypnosis, published in the professional journals, who has attended advanced hypnosis seminars and taught at them, who regularly reads the hypnosis literature, belongs to at least one national professional hypnosis association, and regularly attends and occasionally presents at state and national hypnosis conferences. Such a person has the most in-depth training and experience, the broadest knowledge base which increases with every year of clinical or research experience.

Statistically, the risk of complications would be expected to rise as MQH standards erode, from most to least trained and experienced. In the public marketplace, ads and TV commercials promote hypnosis. Titles such as hypnotherapist, clinical hypnotist, certified, registered, or ethical are used by lay hypnotists to raise credibility and public acceptance. Most often, "certified" refers to a certificate from the individual or organization providing initial training. It has no legal or professional standing. "Registered" frequently means the hypnotist is on a registry or listing of lay hypnotists usually compiled from mailed applications, a "paper review" without examination. A lay hypnotist claimed to be "licensed" and when pressed for details defended this response as having a city retail or business license!

SHOULD THERE BE A LAW?

"As professionals," Kleinhauz and Beran wrote, "our duty is to work to achieve proper legal and medical safeguards against the use of hypnosis by ill trained persons or the entertainment of an audience" (p. 289). Kline (1976) called for restrictive legislation. Marcuse (1964) reported legal constraints to the practice of hypnosis in Australia, Brazil, Denmark, Finland, Great Britain, Italy, The Netherlands, Norway, Russia and Sweden. Bjornstrom (1970) described the 1784 law in France "to put a stop to the scandals of the magnetizers." But he also observed "physicians had neglected to take this process into their own hands and that instead, they had allowed impostors and secret societies to work mischief." In 1815, Emperor Alexander in Russia "declared magnetism to be a very important agency but that just on that account it should be practiced by skillful physicians" (Bjornstrom, 1970). In 1817, the Kings of Demark, Sweden and Prussia limited the practice of "animal magnetism" to physicians.

Where it has been in place, for decades and for centuries, restrictive legislation functions effectively. In the Foreign Letters column of the September 12, 1959 of the Journal of the American Medical Association the following was published:

> When hypnotism is criminal—Paragraph 364 of Norway's penal code provides a fine or imprisonment for whomever illegally uses any procedures whereby he puts someone, even with his own consent, into a state of hypnosis, weakness, or loss of consciousness. The physician who acts in this way in search of knowledge or the cure of disease is exempt . . . it has been invoked against a nonmedical popular entertainer who gave a demonstration of hypnotism in the gymnasium of a high school. The four boys who volunteered to act as experimental subjects responded to the various suggestions, a wand thrown on the floor turning into a snake and an imaginary orchestra playing. When the police instituted legal proceedings against the entertainer . . . a professor of psychology was called as an expert witness. He condemned the performance . . . holding it was likely to do serious harm in the hands of the ignorant. The performer had to pay a fine (p. 206/240).

Ontario, Canada has had restrictive legislation since 1960. Called The Hypnosis Act or Chapter 216, it reads as follows:

> 1. (1) The Minister of Health shall administer and enforce this Act and he may designate any officer of the Department of Health or any medical officer of health or he may appoint any legally qualified medical practitioner for the purpose of making any investigation or inquiry necessary therefor.
>
> (2) Any person designated or appointed under subsection 1 has all the powers of a medical officer of health under The Public Health Act.
>
> 2. Subject to Section 3, no person shall hypnotize or attempt to hypnotize another person.

3. Section 2 does not apply to:
   (a) any legally qualified medical practitioner using hypnosis in the practice of his profession;
   (b) any dentist registered under The Dentistry Act using hypnosis in the practice of his profession;
   (c) any psychologist registered under The Psychologists Registration Act using hypnosis in the practice of his profession on the request of, or in association with a legally qualified medical practitioner;
   (d) any bona fide student registered in a course leading to qualification in one of the professions mentioned in this section practicing hypnosis for the purpose of study under the instruction and supervision of a legally qualified medical practitioner, a dentist registered under The Dentistry Act or a psychologist registered under The Psychologists Registration Act; or
   (e) any member of any class or persons designated by the regulations made under this Act.

4. The Lieutenant Governor in Council may make regulations designating classes of persons to whom Section 2 does not apply and prescribing the terms, conditions and circumstances under which members of any designated class may use hypnosis.

5. Every person who contravenes any of the provisions of this Act is guilty of an offense and on summary conviction is liable for the first offense to a fine no less than $100 and not more than $1,000 or to imprisonment for a term not more than six months, or to both, and for any subsequent offense to a fine of not less than $200 and not more than $2,000 or to imprisonment for a term of not more than nine months, or to both.

6. Every prosecution under this Act shall be commenced within one year from the date of the alleged offense.

In Virginia, an act was approved in March, 1978, repealed in March, 1980 and a revised version offered but failed to be enacted in 1982. The original version approved March 25, 1978 read:

Section 18.2-315. Hypnotism and mesmerism. If any person shall hypnotize or mesmerize or attempt to hypnotize or mesmerize any person, he shall be guilty of a Class 3 misdemeanor. But this section shall not apply to hypnotism or mesmerism performed at the request of the patient by a licensed physician, licensed clinical psychologist, or dentist, or at the request of a licensed physician in the practice of his profession.

The amended version submitted but not enacted in 1982:

Section 18.2-315.1. Hypnotism prohibited. If any person shall hypnotize or attempt to hypnotize any other person, he shall be guilty of Class 3 misdemeanor. This section shall not apply to hypnotism performed by a licensed physician or clinical psychologist in the practice of his profession.

Oregon has had legal restrictions on the use of hypnosis in court since 1977:

136.675 Conditions for use of testimony of persons subjected to hypnosis. If either prosecution or defense in any criminal proceedings in the State of Oregon intends to offer the testimony of any person, including the defendant, who has been subjected to hypnosis, mesmerism or any other form of the exertion of will power or the power of suggestion which is intended to or results in a state of trance, sleep or entire or partial unconsciousness relating to the subject matter of the proposed testimony, performed by any person, it shall be a condition of the use of such testimony that the entire procedure be recorded either on videotape or any mechanical recording device. The unabridged videotape or mechanical recording shall be made available to the other party or parties in accordance with ORS 135.805 to 135.873.

136.685 Law enforcement personnel required to advise hypnosis subjects of consequences; consent of subject required.

(1) No person employed or engaged in any capacity by or on behalf of any state or local law enforcement agency shall use upon another person any form of hypnotism, mesmerism or any other form of the exertion of will power or the power of suggestion which is intended to or results in a state of trance, sleep or entire or partial unconsciousness without first explaining to the intended subject that:

  (a) He is free to refuse to be subject to the processes delineated in this section;

  (b) There is a risk of psychological side effects resulting from the process;

  (c) If he agrees to be subject to such processes, it is possible that the process will reveal emotions or information of which he is not consciously aware and which he may wish to keep private; and

  (d) He may request that the process be conducted by a licensed medical doctor or a licensed psychologist, at no cost to himself.

(2) In the event that the prospective subject refuses to consent, none
of the processes delineated in subsection (1) of this section shall
be used upon that person.

The international standard of hypnosis legislation is to limit its practice to licensed physicians, psychiatrists, psychologists, and dentists who
have training and experience in the use of hypnosis (Bjornstrom, 1970;
Marcuse, 1964). Very few states, provinces and nations hypnosis laws
allow social workers and nurses the independent, unsupervised practice
of hypnosis. Readers who are interested or involved in introducing proposed restrictive legislation should consider this checklist of content:

1. Identify **what is covered** (hypnosis).
2. Name the **professional categories** and **specify independent level
   of practice** (licensed physicians, clinical psychologists, dentists);
   specify **what is forbidden** (radio, television, filmed, taped, public
   entertainment inductions).
3. Name **others who can practice under supervision** of the above
   (nurse anesthetists, Master's level psychologists, clinical social
   workers, etc); **define and specify supervision** (should be based
   on written agreement, required hours, documentation).
4. **Informed consent** must be obtained from subjects.
5. Subjects must be told of **benefits reasonably expected** and **the
   risk potential.**
6. Subject must be given **freedom to discontinue** at any time.
7. **Forsenic guidelines** (entire sessions must be videotaped; subject
   can discontinue "without prejudice").
8. **Penalties:** violations considered a misdemeanor; fines, imprisonment.

LICENSING VS. SELF-REGULATION

If a state, province or nation does not have legislation restricting the
use of hypnosis, would it reduce the risk of complications to require that
hypnotists be licensed? To do so would be to establish yet another separate grouping of practitioners, adding to an already confusing assort-
ment of service providers at a time when legislatures are sunsetting
licensing boards or threatening to do so. Medicine, dentistry and psychology clearly state through their national associations that hypnosis is
an adjunct to their professions, an additional method and
technique and not a profession itself.

Some argue that licensing does not ensure competence. Gross (1984) is critical of the licensure model. He feels it is really a form of professional self-regulation, not public regulation, aimed at creating a relative monopoly of practitioners who meet unvalidated, arbitrary, self-imposed standards (e.g., academic achievement; currently popular theoretical orientation; test-taking recall and ability; self-approved curricula and internships, etc.). He claims "success in schooling only predicts success in schooling, not competence or effectiveness" and competency studies do not differentiate the licensed from the unlicensed. State licensure considers competence a function of educational qualifications, continuing education and monitoring by ethics and peer review committees, licensing boards, and to a very minimal degree, the public.

Gross argues that there is no evidence the type or amount of professional education or continuing education protects the public or ensures competence. Ethics and peer review committees and licensing authorities are all largely ineffectual, investigating and disciplining few impaired, incompetent or exploitive practitioners and when they do so there are few heavy penalties. He attributes these to vague evaluative criteria, fear of litigation, avoidance or practitioners to criticize peers, and "group solidarity and secrecy." Emphasis is more process (activities) and structure (rules and regulations) than on outcome (product of services). Licensing, according to Gross, is more a benefit to the profession than to the public.

He recommends more consumer involvement in professional credentialling, giving the public more responsibility for their own self-protection. He feels professional service providers should function more as partners with consumers, as advisers and educators rather than decisionmakers or mystifiers. In short, professionals should become more accountable. He suggests that the rise of consumerism indicates a commensurate rising sophistication. This, in turn, causes society to be more demanding of quality services and of more information from which services can be compared. The rise in malpractice cases (and insurance rates) and the current state of forensic hypnosis supports Gross' views. If professionals cannot satisfy public demand, eventually the courts will take action to protect the public interest.

HIGH RISK HYPNOTISTS, SUBJECTS AND SETTINGS

From the complications described in earlier chapters, a worldwide sampling, historical and contemporary, in clinical, research and entertainment settings, it is possible to describe generally stereotypes of high risk hypnotists, subjects and physical settings:

CRH (Clinical Risk Hypnotist) would not likely be a licensed physician, psychiatrist, psychologist or dentist, would have fewer years of formal education, professional training and supervised experience, fewer hours of hypnosis training and little or no supervised hypnosis experience. CRH attended one or two hypnosis workshops, years ago and hasn't taken further continuing education in hypnosis. There was no written, oral or practice examinations to establish competence. CRH is likely to be a "disciple" of one theorist or method and uses it exclusively. Very little history is taken. There is no informed consent. Subjects are asked what they know about hypnosis and are told only about what they ask. Few notes are taken on the subject's questions, concerns behaviors or responses. CRH often sits with eyes closed and lights dim, unable to clearly see the subject. CRH "never does the same thing twice" and "experiments, will try anything once." Dehypnosis, neutralizing posthypnotic suggestion, amnesia or cues (i.e., white sound) are short, superficial or not at all. There is no debriefing or re-entry, subjects are asked casually "how they're doing" and most reply equally casually: "Fine!" or "More relaxed."

CRS (Clinical Risk Subjects) are generally more concrete than abstract, have difficulty with indirect methods and have constricted affect usually with a tendency to overcontrol emotions. Often there is repressed hostility toward authority figures, expressed in passive or passive-aggressive reaction formation. There are several CRS variations. Some have histrionic traits or are loose, tangential or circumstantial to near psychotic proportions. Others are more obsessive-compulsive, and still others have strong dependency needs.

RRS (Research Risk Subjects) are similar to CRS, their brothers and sisters, but with a high degree of inadequacy, inferiority feelings which can signal an identity crisis. They may be homesick, further reinforcing dependency needs. SRS (Stage Risk Subjects) combine all the "family traits" or risks of their siblings and peers, CRS and RRS.

RRH (Research Risk Hypnotists) are so obsessed with their experimental design and its methodology that subjects are "run" through the experiment like rats through a maze. In many cases, RRH is not consciously aware of this high-risk attitude. Informed consent is largely a clerical, administrative task. There is little discussion or conversation with subjects. Instructions are read in a monotone, RRH seldom looking up from the typed sheets ("saves time, more objective, scientific, standard lab conditions"). Dehypnosis is less important than debriefing "because it isn't specified in the protocol." RRH conducts experiments most any place at any time convenient to him/her and as quickly as possible, acquiring the most data in the least time. The subject's mood, readiness, comfort, convenience or mental state are secondary considerations "not really part of the experiment." Followup is available through the college counseling center or local mental health clinic: "I give 'em a card before they leave in case they have any trouble."

SRH **(Stage Risk Hypnotists)** are interested in "putting on a good show in the spirit of fun." There is not enough time for screening and subject selection and only the best (most hypnotizable) subjects are selected as quickly as possible. Those not selected return to the audience without dehypnosis, debriefing or further observation. The "good" subjects "make up the show," regressed to infancy or early childhood or confront hallucinated tastes, smells or embarrassing social situations. The more dramatic and emotional they react the more effective the "performance." The more "magical" the show or charismatic the hypnotist, the greater the box-office receipts. As Shakespeare put it: "The play's the thing."

Values differ markedly among hypnotists according to the setting. Clinical hypnotists tend to focus more on individual differences, personal factors, to most effectively help with specific needs. Researchers can be sensitive socially but their prime emphasis is the pursuit of truth. Working with and through people is a means to the end. Stage hypnotists can also have a social or public conscience but the entertainment program, "the show," is a sequence of rapid inductions with little time for adequate screening and history taking, minimal observation, or dehypnosis and debriefing.

A stereotype of a high-risk physical setting would be, for the specific subject, the most uncomfortable atmosphere and environment. It is unlikely these factors alone would cause complications but in combination with subject and hypnotist risk factors, they can be the proverbial "straw that broke the camel's back." Poor acoustics, a bright light source in the subject's eyes or face, and a hard, straight-back chair would very likely be less comfortable and less conducive to relaxation and the hypnotic process than a comfortable recliner chair with subdued light. Combine an uncomfortable, unpleasant environment with high risk subject and hypnotist factors and the probability of hypnosis complications is increased.

THE NEED FOR IMPROVED TRAINING

The 1962 position paper *Medical uses of hypnosis* (Group for the Advancement of Psychiatry) recommended that:

> Whoever makes use of hypnotic techniques . . . should have sufficient knowledge of psychology, and particularly psychodynamics, to avoid its use in clinical situations where it is contraindicated or even dangerous. Although similar dangers attend the improper or inept use of all other aspects of the doctor-patient relationship, the nature of hypnosis renders its inappropriate use particularly hazardous . . . more than a superficial knowledge of the dynamics of human motivation is essential" (p. 705).

That report also contained specific recommendations for training programs, among which was the following: "Isolated courses limited to the teaching of trance induction techniques are strongly disapproved" (p. 705). There has been considerable improvement in professional hypnosis courses since these 1962 recommendations but other areas of need have emerged:

Most hypnosis training is generic, the same course for physicians, dentists, psychologists, and increasingly social workers, Master's degree psychologist, and registered nurses. Few courses are designed and conducted by or for specific professions. Hypnosis training in The Netherlands, however, offers a 56-hour introductory course restricted to licensed physicians, psychologists and dentists in groups of 12, meeting weekends every six weeks. This is followed by advanced courses of 18 days, meeting one weekend a month, for psychiatrists and psychologists. Physicians and dentists take an advanced course of only eight days (Rostafinski and MacHovec, 1986).

Hypnosis training in Europe is based on a slower, more gradual skills development. Italy offers a 120-hour course restricted to physicians, dentists and psychologists over a three year period, 40 hours per year. Germany provides a 10-hour introductory course, then an advance course of 40 hours in four 1-½ day sessions over a 6-month period, to physicians, dentists and psychologists. European hypnosis training is based on a more gradual development of skills, from six months (Germany) to three years (Italy), contrasted to the American "fast food" model of one or more weekend sessions at much closer intervals, usually consecutive weekends.

Can such "instant skill" courses contribute to the risk of complications? There is no empirical data at this writing but Rostafinski and

MacHovec (1986) polled 900 graduates of an introductory course by one of the established national hypnosis associations and respondents frequently reported they "felt overconfident" by the training. Taking more time in training would provide more gradual skills development and a more careful level of practice. A powerful contributing factor to hypnotist over-confidence is the lack of emphasis on risk factors and what can go wrong. If you are taught hypnosis is "safe" you will be less aware of danger, less cautious.

Worldwide, few training programs provide a block of time or lesson plan specific to hypnosis complications and risk management. Graduate hypnotists are therefore relatively unaware of risk situations, what to look for and what to do. There is a very real need to include specific information on subject and hypnotist risk factors and preventive practices for subject safety and the hypnotist's legal protection. There is a need for a solid, basic introductory course, at the level of sophistication of the licensed physicians, clinical psychologists and dentists attending and to continue at that same level to develop mastery of the subject, seasoned judgment and optimal use of intervention skills.

Graduates of the basic course should be assigned mentors, a combination case consultant and personal supervisor, as they proceed through specialized training in their profession. Ideally, mentors should be hypnosis diplomates by their respective Board. The 1962 position paper emphasized the need for close supervision: "although lectures, demonstrations, seminars, conferences and discussions are helpful, the basic learning experience must derive from closely supervised clinical contact with patients . . . in their particular field of medical competence" (p. 706).

INFORMED CONSENT

Despite the many malpractice suits nationwide and the resulting increase of insurance costs, many hypnotists do not feel legal informed consent is necessary in the use of hypnosis. Few professional hypnosis organizations specifically recommend that informed consent be obtained for clinical purposes. It is the author's impression that informed consent is obtained far more in experimental research, due to stringent requirements for research with human subjects. Written informed consent is almost never obtained for stage hypnosis.

Baumann (1985) stated: "I believe it is entirely ethical to use hypnotic approaches without first telling the patient." He referred to Milton

Erickson, "one of the most respected of all hypnotherapists" who induced hypnosis in patients "without their knowing what he was doing." Baumann compares the uninformed use of hypnosis with a surgeon who "must tell the patient that he intends to remove an appendix" but "need not specify that he intends to use hemostats, Allis clamps, or other specific tools of his trade. The psychotherapist need not inform the patient that he intends to use interpretation, clarification, or behavior modification techniques. I feel that the same is true regarding hypnosis" (p. 414).

He considered the possibility that a patient might sue a hypnotist, claiming "he was hypnotized without his consent and that he continues to feel that he is under the influence of the hypnotherapist." The reader will note that several cases in this book fall in this category. "I believe," Baumann concluded, "most experts would agree that the hypnosis was not the source of his feelings of being influenced, but the ways of God and the courts are mysterious to behold" (p. 414-415). Hare-Mustin et al. (1979) observed that "some psychoanalytic therapists may feel that detailed information about therapy and the therapist weakens the transference relationship" (p. 7).

Hypnotists who assume they will never be sued and that expert witnesses will convince a court that informed consent is not necessary may have a rude awakening. The fact is that courts are increasingly deciding informed consent is necessary for any therapeutic intervention. It is already required routinely for research purposes. What follows is a description of the legal, ethical and philosophical bases of the informed consent doctrine.

Informed consent is a mutual agreement free of coercion or deceit between an informed, competent consumer and a professional provider competent to deliver the services needed. It is a legal mandate rooted in English Common Law and the responsibility of the service provider. It cannot be abrogated or relinquished. Hare-Mustin et al. (1979) stated that informed consent is "developing a contractual agreement" by "negotiation between partners" (p. 7).

Legal principle. If a practitioner does not provide sufficient, appropriate information to a consumer before performing a specialized service, the consumer may litigate for damages if injury is sustained even if the service provided was properly performed.

This is also an **ethical and philosophical principle.** The service provider and consumer join in the inherent duty and right of one human being to proceed in good faith to do good and avoid harm to a fellow human being. The intent is to provide substantive benefit to the consumer and also to society. Informed consent is in the Western tradition of ensuring personal autonomy, free choice, and the dignity and integrity of the individual.

Lidz et al. (1984) described three "classical elements" of informed consent:

1. **Nature of the procedure,** whether diagnostic, therapeutic or experimental, the agreed purpose, and benefit(s) reasonably expected;
2. **Alternative treatment and providers,** the benefits and risks of these or of no treatment at all;
3. **Risks.** "Although disclosure of the risks neither guarantees that the patient will use the information making a decision nor assures that the decision reached will be a reasonable one, patients cannot make an informed decision without knowledge of the risks" (p. 12).

Hare-Mustin et al. (1979) referred to two factors which should motivate therapists and researchers to obtain informed consent: rising consumerism emphasizing the rights of consumers, and the great rise in malpractice litigation. The increased emphasis on client rights requires more client responsibility: "Clients are expected to choose wisely, to make use of information provided, and to assume control of their participation in therapy" (p. 4). They describe three client needs:

1. "Clients need **knowledge of procedures, goals and side effects** of therapy . . . information about possible indirect effects is essential if the client is to weigh the benefits and risks of entering treatment" (p. 5);
2. Clients need information about the **qualifications of the therapist.** "Providing a description of skills and experience and responding to clients' queries safeguard both therapist and client from unrealistic expectations the client might hold" (p. 6);
3. Clients need to know about **alternatives,** the local network of available services and service providers.

During the 1970s there was a change in malpractice case law from "the professional community standard" to that of "the reasonable man." The professional community standard is what professional colleagues or peers of the service provider consider appropriate disclosure, regardless of whether or not the consumer understands the information. Under this standard, the practitioner determines what kind of information and how much of it is provided. Some courts have ruled disclosure should be to the extent it is "material" for the consumer to give or deny consent, as judged by practitioners and not consumers. If this standard applies, information considered customary and usual is defined by expert witness testimony.

In 1972, in *Canterbury v. Spence,* the reasonable man standard was introduced, a consumer-oriented criterion in tune with the emerging emphasis on individual self-determination, the right to understand, consider or decline the service offered. This presumes that the average citizen in a free society has the ability as well as the right to weigh advantages and disadvantages and arrive at a rational decision in his or her best interests.

There is debate over the objectivity of informed consent in these two standards. Does the reasonable person refer to the specific person or humanity at large? If it is the individual is this not a subjective measure? If it is not, how does the court judge reasonableness of society? Who can best determine the adequacy and appropriateness of information to be provided, the practitioner or the consumer? Some courts have not differentiated between these two conflicting standards but increasingly the reasonable person standard is emerging. Lidz et al. (1979) sum up this dilemma:

> The dust from the storm that this issue has created has not yet settled, and it is unlikely that one rule will emerge to the exclusion of the other. Rather, these two different standards . . . probably will remain, with some jurisdictions subscribing to one and the remainder to the other. Despite the problems inherent in the patient-oriented standard, primarily that of how a physician is to know what a reasonable patient would want to know, it is the preferable standard . . . because it is most in keeping with the values underlying the informed consent doctrine and the goals that the doctrine seeks to promote, especially that of assuring the patient's primacy in decisionmaking (p. 14).

From the foregoing, it would seem advisable from a risk management standpoint to satisfy both disclosure standards. In this way, hypnotists can demonstrate they satisfied appropriate standards of care or the "state of the art" as substantiated by expert witness testimony from their professional colleagues. At the same time, they can document evidence of informed consent from subjects who were given information which would be understandable to a reasonable person. To do neither is to run the risk of being defenseless if ever a subject litigates for malpractice. The following is a suggested informed consent procedural checklist:

1. **Information,** in clear, simple language, free of jargon, of what hypnosis is (not yet defined; current major theories; all hypnosis is

self hypnosis); what hypnosis is not (magic, mind control, common misconceptions); the technique to be used.

This is the same ethical requirement of physicians, dentists ad psychologists with respect to medical, dental or psychological tests and procedures. Everything known by the practitioner need not be disclosed if that information is judged by the practitioner to be too complex to be understood by the average lay person. As one court pointed out, for a physician to tell everything s/he knows would be equivalent to the patient going to medical school.

Insufficient, inadequate or confusing information is poor informed consent but does not constitute negligence unless there is proven harm or injury. Even then it must be established the injury was the result of the lack of informed consent. Some argue a lack of consent of and by itself constitutes harm and injury because the person's dignity has been violated even without experiencing complications. It is not likely such cases would succeed in court unless lack of consent can be shown to contribute to or cause personal injury.

The potentially strong symbolism or imagery of the hypnotist has risk potential even when providing information. Hare-Mustin et al. (1979) described several variations just in describing alternative treatment: "Therapists could be cautious to characterize alternative systems fairly" implying a value judgment against them or "when a course of action is insisted upon . . . the client's right to free choice is diminished . . . a premature or overly enthusiastic discussion . . . may convince the client that the therapist does not want him or her in therapy."

Lidz et al. (1984) recommend giving a copy of the signed informed consent agreement to the subject and also written background information "long in advance of their actually being requested to make a decision" (p. 331). The author routinely mails a descriptive brochure of background information together with questionnaires of historical information to all prospective clients (see Appendices).

2. **Benefits** which are reasonable to expect (no guarantee)
3. **Risks** (by percentile incidence, most frequent named; assurance of prompt intervention and treatment or referral to another professional if preferred)
4. **Alternatives** (other services; other providers; benefits and risks). A sampling of typical alternatives: community mental health

clinic; private clinics; physicians and medical treatment (medications); psychiatrists; social workers; pastoral counseling; vocational or rehabilitative counselors; peer support groups (AA, Al-anon, weight watchers, parents without partners, parent effectiveness training); assertiveness training; biofeedback; meditation; crisis centers or hotlines; day centers; sheltered workshops; legal assistance. Local alternative services could be printed and included in an "informed consent packet."

5. Opportunity to **ask questions and have them answered** in clear, understandable language.
6. **Freedom of choice** (professional community and reasonable person standards/ "free agent" to accept or decline treatment; freedom to stop treatment or experiment at any time).
7. **Signed consent form.** It may come as a surprise that courts have held that a signed, printed informed consent form does not entirely satisfy legal requirements. It is expected that practitioners document the discussion or conversation with the subject (e.g., progress notes, notation of any questions asked, date and time listed). Lidz et al. (1984) recommend that "consent forms that provide minimal information and a place for the patient's signature may be preferable to forms giving details of a procedure, since the former are less likely to be used as a substitute for discussion" (p. 331).

Items 1 through 5 can easily be incorporated in printed material given or mailed to subjects before the first hypnosis session. The appendices following this chapter contain samples of checklists used by the author since 1981 which help ensure these important legal aspects are discussed. Clinicians and researchers might find it convenient to combine informed consent with a written agreement or contract for the therapy or research which is to take place. If so, the combined form should include all of the points discussed above plus these:

1. Description of mutual reasonable expectations of the subject and the clinician or researcher.
2. Definition of the subject's self-responsibility to assure informed consent and free choice.
3. Goal or purpose of services to be provided with outcome probabilities described.
4. Probable number and length of sessions required.
5. Fees and method of payment.

6. Subject access to resulting written records.
7. Terms of termination, transfer, renewal or revision.

ETHICAL CONSIDERATIONS

Ethical principles for clinical and experimental hypnosis shared by physicians, psychologists and dentists would help establish multidisciplinary standards of care and also facilitate informed consent. The author has used the following Code of Ethics printed on the reverse side of the informed consent form and also prominently displayed in the waiting room:

I RESPECT YOU as an individual,
   with dignity and integrity,
   morals, values and standards,
   a unique personality,
I will do nothing to harm you.

YOU HAVE RIGHTS
   to accept or decline treatment,
   to stop treatment at any time,
   to ask questions and get answers
I respect your rights.

OURS is an ethical, professional relationship,
   sensitive and confidential,
   without deception or deceit.
We will do only what we agree to do.

I HAVE AN OBLIGATION
   to know about you (take a history),
   to explain what we will do and
   how we will do it,
   to explain later what we've done,
   to take care of you,
to protect your mental health.

Psychiatrists, psychologists and dentists are bound by their own ethical principles and standards of professional practice but do not have a separate code of ethics expressly for hypnosis. Professional hypnosis associations such as the International Hypnosis Society, Society for Clinical and Experimental Hypnosis, American Society for Clinical Hypnosis, each have their own ethical standards. The following is a distillation of their conceptual content:

1. Practice of hypnosis should be **limited to areas of competence** established by knowledge skills and judgment from professional education, training and experience and in addition appropriate training and supervised experience in hypnosis by fellow professionals themselves trained and experienced in hypnosis.

2. Training and supervised experience shall include an **adequate knowledge base** of facts and theories, **risks and limitations, variety of techniques and applications,** according to **current standards** of professional education in medicine, psychology and dentistry, with **continuing education** for ongoing skills development.

3. Practitioners place the highest priority on **caring for the individual subject** who, by consenting to treatment or as a research subject, places trust in hypnosis and the hypnotist. This caring relationship is evidenced by:

   Appropriate informed consent consisting of
   adequate information clearly presented;
   benefits reasonably expected;
   risks named with percentile incidence;
   alternative treatment, benefits and risks;
   opportunity to ask questions;
   freedom to accept or decline
   and to stop at any time;
   signed informed consent form.

   Careful observation
   before
   during
   after induction

   Dehypnosis
   adequate and thorough
   or rehypnosis if needed

   Debriefing and re-entry
   to protect or restore
   prehypnotic mental state

   Followup or referral
   if and as needed

   Confidentiality
   appropriate to the situation

4. Restriction to the use of hypnosis:

   Limitations explained and observed
   (age regression or progression;
   hypnotically-enhanced recall)

Never for private or public entertainment or demonstration
(except as part of professional
hypnosis training in appropriate setting)
Never for personal gain
(to deceive or defraud
exploit or manipulate)
Never to teach lay hypnotists
(those not licensed physicians,
dentists or psychologists)

Malpractice

Malpractice cases are **civil** not criminal actions and are classified as **tort law.** The **plaintiff** (injured party) seeks "relief" from the "wrongs" of the **defendant** (you!). Relief can be **nominal, compensatory** or **punitive.** Compensatory damages are usually based on lost wages or pain and suffering. Punitive damages are "exemplary," seeking to hold you up as an example and punish you. Wrongs are classified as **intentional, reckless,** or **negligent. Reckless wrongdoing** is a conscious disregard of known risk. **Negligence** is a departure from accepted practices, from standards of care, acting unreasonably or lacking in skill. Evidence in malpractice trials is the **preponderance rule** or 51% certainty, better than chance. In criminal law the rule is **beyond a reasonable doubt,** or 85%, that a reasonable person would not disagree with the finding. In civil commitment hearings, the rule is **clear and convincing evidence,** or 75%.

There is a **pretrial review** to determine whether the case will go to court and at this stage the burden of proof is on the plaintiff. If the case goes to court, the burden of proof then changes to the defendant. If the disclosure standard is the professional community, the defendant presents expert testimony to demonstrate there were adequate standards of care. If the standard is reasonable person, testimony will be presented to satisfy that criterion. The most likely grounds for malpractice suits against hypnotists would be:

Faulty diagnosis. These would be the "horror stories" of misdiagnosed headache for a brain tumor, anxious subject who suicides or decompensates into psychosis, etc. An effective legal defense in such cases is to offer expert testimony that another well-trained or similarly trained hypnotist could also have missed the diagnosis.

Lack of informed consent, where severe complications occur and the subject litigates on the basis there was inadequate information provided as to risk, that the hypnotist could or should have known that the complication could occur and took no action or was incompetent to know what to do.

Abandonment, where a subject does not show up for a session and is not contacted, reminded or offered another appointment and suicides or suffers severe complications and survivors sue on grounds similar to those for lack of informed consent.

Breach of confidentiality or **invasion of privacy,** where there is no reason to divulge information, such as to a subject's employer, spouse, friend, neighbor, or other without a signed release of information or the subject's knowledge or consent (no one in imminent danger).

Sexual contact or conduct, where hypnosis was allegedly used to seduce a subject or so manipulate or influence them as to enable the hypnotist to take sexual liberties.

Wrongful death. The usual legal criterion here is whether or not the subject's death was foreseeable (previous history, hypnotist's observations and judgment). While it may be unlikely a hypnotist would be charged with wrongful death; an otherwise unexplained suicide or accidental death shortly after a hypnosis session could conceivably result in such a charge.

A plaintiff must prove five points in order to make a malpractice claim actionable:

1. There must be a **professional relationship.**
2. Duty to the client required **skill and care.**
3. That **duty was breached** (by preponderant evidence).
4. And an **injury was sustained.**
5. The **proximate cause** of injury **was the breach** of duty (or if there was no breach there'd be no injury—called the **"but for . . ."** test).

It is difficult to prevent being charged with malpractice. Certainly, the longer one is in practice and the larger that practice is, there are increased odds someone will sue. The best preventive practice is:

To take a **careful, detailed history;**
Obtain **adequate informed consent;**
Carefully observe the subject before, during and after trance;
Document your awareness of any potential risk

Observe duty to warn if the subject names a potential victim and has the means and a violent history

If in doubt about a subject's behavior **consult a colleague,** discuss it

Keep **adequate records** to document a good faith effort and reasonable effort to reduce risk

Follow **standard procedures** of **a checklist sequence** of **adequate standards or care** consistent with your profession.

RISK MANAGEMENT

Risk management is the overall strategy for preventing or reducing hypnosis complications. The subjects presented earlier in this chapter describe underlying legal, professional and ethical components of an effective risk management system. The purpose is to ensure public safety and welfare and to assure the public and the courts that there is accountability. The cornerstone of accountability is documentation, written records which clearly demonstrate conscientious, good faith effort to provide the highest quality of services at the most competent level of practice. Even if it can be proven in court that a complication was inevitable, it must also be proven that the practitioner did nothing to cause harm by negligence or lack of care or skill, that the complication would have occurred to any other hypnotist.

The hypnotist's standard or routine procedures which are the same for all subjects should be documented routinely. While the major reason for this is to protect the subject, printed forms, checklists and written notes are documentary evidence of the existence of adequate standards of care, of great help in any malpractice litigation, more easily defended by expert witness testimony of colleagues. Since 1970, the author has developed a documented system of hypnosis to minimize subject and hypnotist risk factors. It helps isolate and identify potential problem areas or behaviors which, because they are known before, can be avoided, neutralized or used in hypnosis. Called "the envelope" it is an 8-step system of risk management:

1. **History**

   As Walker (1967) emphasized "the prime tool for the thorough understanding of preventing psychiatric signs and symptoms (whether on a psychological or somatic basis) is the scrupulously taken history" (p. 7). The major obstacle to taking a careful history is the time it requires. Some clinicians and researchers take the time to do so. The author knows several who use a 2-hour block of time to interview subjects for therapy or research. This largely administrative task encroaches on therapy or research time and cannot be delegated because the information obtained is important.

Procedure: subjects who phone for information regarding hypnosis are mailed a brochure which describes hypnosis and the hypnotist's qualifications, a form listing basic subject information (name, address, etc), on the reverse side of this sheet a hypnotizability checklist the author created from susceptibility scales, clinical and experimental literature and his own experience, and a 2-sided sheet of medical history on one side and a checklist of anxiety, fears or phobias on the other. Thus, the prospective subject gets three pieces in the mail:

- A description brochure (four 8½ x 11₀ sheets folded in half, center stapled);
- An "application" type of preliminary demographic information with hypnotizability checklist on the other side (Appendix A this book);
- History checklists, medical on one side, sources of anxiety and/or fears on the reverse side (Appendix B this book).

Documentation: Phone calls are logged by date, time and the person who called. The date material is mailed is also entered.

2. Briefing, Informed Consent, Contract.

When the subject arrives for the first appointment, the completed information and history forms have been received and reviewed. The subject is met and rapport established more easily than when the information must be obtained, reviewed and applied "on the spot." At this time, the hypnotist has the file of papers received and an induction and treatment summary form on which key points are written which are directly related to hypnosis (see Appendix D this book). Full legal informed consent is then obtained, orally and in writing. The author combines consent and a treatment contract in the same form, with his own "Code of Ethics" on the reverse side.

Procedure: Mailed history and information forms are reviewed and provide the basis for informed consent/treatment contract, orally and in writing (Appendix C). The informed consent form is used as the basis for a structured interview, discussed point by point, providing an overview of the hypnotic process (what hypnosis is and is not, benefit-risk aspects, alternatives, etc.).

Documentation: Information and history forms are placed in the subject's file together with the original of the signed and dated informed consent form (subject gets copy) and hypnotist shifts attention to the induction/treatment form on which is noted important individualized needs (purpose of treatment, subject's own choice of most pleasant visual imagery and that which should be avoided, hypnotist's own notes of areas for special attention).

3. **The Standard Induction.**

The induction/treatment summary, on a clipboard in the hypnotist's hands (and with pen/pencil at the ready), is used for a quick second briefing, point by point following the checklist sequence on the sheet (see Appendix D).

Procedure: The subject's informed consent is verbally assured (given opportunity to ask questions) and made optimally comfortable (best recliner chair position, pillow, blanket, lights dimmed and reason for doing so explained). Every subject is given the same standard induction, consisting of physical relaxation with a "long count" (from 30), followed by suggested visual imagery enhanced with background white sound (subject chooses ocean, rain or waterfall, volume and tone) and a "short count" (from 10) suggesting increased relaxation and more vivid visual imagery.

Documentation: Hypnotist refers to induction/treatment summary and makes notes on subject's observed behaviors and responses to the induction.

4. **The Work** (Contents of the Envelope, the Message).

This is the individualized message to the subject. For clinicians it is the motivating or explorational message based on agreed therapeutic goals. For researchers it is the "meat" of the experiment, the means to the end of the research objective.

Procedure: Hypnotist at this point is on her/his own and pilots the ship of hypnosis by hand, custom fitting every word and thought to the unique needs of the individual subject, or for the researcher, sharp focus on the experimental needs. The hypnotist carefully observes the subject at all times, aware of the verbal and nonverbal signs and aware of the hypnotist's own verbal and non-verbal behaviors.

Documentation: Hypnotist makes notes as appropriate on the induction/treatment summary or if in therapy or with special research goals, on a separate sheet or progress note.

5. Dehypnosis

The "message in the envelope" having been "read," the hypnotist now guides the subject back to the pre-induction state, suggesting recall as appropriate. Treatment goals in therapy may require more or less recall or a graduated recall of what was experienced during trance.

Procedure: If white sound or any other sound effects were used, these are specifically dehypnotized to prevent spontaneous trance or partial trance if the subject hears similar sounds outside the treatment or research setting. Any subject, event, thought or feeling which might be harmful or embarrassing to the subject is neutralized or blocked with hypnotic amnesia (and noted in writing). If hypnosis is to be repeated, the suggestion is made that the subject will find it easier to relax at the next session.

Fixing the Seal. It is at this point that every subject is given the posthypnotic suggestion: "No one will ever be able to hypnotize you unless you want to be or need to be, unless you know and trust the person who is doing it and know that that person is qualified. And so, no one will ever be able to hypnotize you without your knowledge and consent. This is for your own safety and protection." This is a very old practice, of unknown origin, but reported by several authors, the oldest Wolff (1936) and Teitelbaum (1965). Milton Erickson would say to his subjects: "You don't have to listen to me" and he reported that in some cases they didn't! He would also say: "Nobody can control you. You can defy me any time you want to, or anybody else. You are a free citizen and be free with yourself" (Erickson and Rossi, 1981, p. 232). The author does not recommend suggesting to subjects "nobody can control you" or the word "defy." These words can, in some, sensitize them to control issues, their losing control, the hypnotist taking control and the subject's need to "defy" or "submit." The author's wording is suggested as involving less risk.

Meares (1960) "fixed the seal" though he did not refer to it as such: "Patients who are easily hypnotized are given the suggestion that they will never go into hypnosis except for a physician or dentist, and only if they wish to themselves" (p. 91). Meares describes, as did Erickson, that some of his patients were unhypnotizable "a few sessions later." In sixteen years of routine "fixing the seal," the author has never had a subject who was not later hypnotizable using the wording given earlier in this chapter. But many subjects have reported, usually with a broad smile, that they have been immediately aware of any salesman, politician or religious zealot who "came on" to them in an attempt to influence their behavior. Even if "fixing the seal" is not 100% effective, even if it can be circumvented by cleverly manipulated techinque, it is in keeping with rising consumerism, client rights, and the emerging emphasis on a free citizen's right to self-determination. It is also a good faith effort to protect the subject's well being and mental health, consistent with the ethical principles of all professions. The author recommends it as standard practice.

Documentation: Hypnotist makes pertinent notes on the induction/treatment summary.

6. Debriefing.

At this stage, the subject describes in his or her own words the subjective experience of hypnosis in an unhurried atmosphere. The subject's thoughts and feelings are compared and contrasted before, during and after hypnosis and at the present moment, with the hypnotist attentive to any deviations (e.g., invading thoughts, physical or mental symptoms, recall of early life events not part of the experiment or therapeutic goals, etc.).

Procedures: informal discussion of the subject's experiencing of hypnosis. This is a separate and distinct phase of the hypnotic process which recent research on eyewitness recall has shown to be an important variable, affecting subject memory. Since memories of significant life events are active personality determinants, these few minutes of informal sharing are a critically important to effective, ethical hypnosis.

Documentation: Hypnotist notes any pertinent information.

7. Re-entry.

This is not at all the same as dehypnosis or debriefing which are parts of the hypnosis process and involved directly in it. Re-entry is preparing the previously hypnotized subject to re-enter his or her life situation. For most subjects, re-entry can be accomplished simply by sitting a few minutes (half hour is recommended) in the waiting room after hypnosis so that thought processes and reflexes can return to "normal" for the drive home or the return to work.

Procedure: After debriefing, the subject is walked to the waiting room and asked to remain there for 20-30 minutes. The author has found that subjects respond favorably to this since it reflects the hypnotist's caring attitude. Hopefully, this is the genuine attitude of the hypnotist but it also provides an opportunity to detect after effects which might emerge while it is still early to control and eliminate them.

Documentation: The hypnotist should note the time the subject finally left the waiting room. For readers who feel these procedures are unduly compulsive or unnecessary, it is pointed out that this note documents a complete written record of standards of care from first phone call to last contact for every subject treated or participating in research.

8. Followup.

This is an optional step, taken if ever a subject does not keep an appointment or phones or writes to cancel an appointment or terminate therapy or an experiment. If step is not taken, it opens the door for possible charges of "abandonment." There have been suits filed on the basis of a subject's initial phone call to a therapist, despite the fact that no appointment was made and the subject was completely unknown to the therapist.

Procedure: If there has been any sign of unusual response to hypnosis, or in every case where a subject for any reason does not keep an appointment or terminates therapy or participation in a research project, a phone call should be made and in addition a letter written offering another appointment or specifying an alternative.

Documentation: A written notation is made in the subject's record of the phone call. A copy of the letter is placed in the file.

Risk management is, in a sense a three-handed partnership of hypnotist, subject and society at large (the law, the courts, legislators, consumer advocates). The single most effective way to demonstrate good faith is by written documentation. It may be inconvenient but it is irrefutable. The author has found that once a risk management system is in place, it becomes routine and involves very little more time than that the hypnotist should have been devoting anyway. Most good practice is good law, which is also good common sense.

PREVENTIVE PRACTICES, CRISIS INTERVENTION

If risk management is strategy, then **preventive practices** are tactics. They are, actually, the same intervention skills most experienced psychotherapists learn either by study or "the school of hard knocks," practical experience. Preventive practices are those used before a complication arises, based on anticipation and an appraisal of possible risks which might develop. An example of this would be a male hypnotist who proceeds with care treating a lonely, recently divorced woman his own age, or a female hypnotist treating a male her own age who feels isolated and unwanted. Crisis interventions are used "in the heat of battle" when the hypnotist is involved in moderate to severe complications which suddenly erupt with great force during the hypnosis session.

Before considering these tactical interventions, hypnotists should review the following as a pre-test of their own readiness and preparation to deal with emergency situations:

1. Effective Prevention = Effective Preparation. Objectively and dispassionately review the quality and recency of previous hypnosis training and experience according to the content dealing with complications and preventive practices. Are you competent to quickly identify, promptly intervene and effectively treat the complications described in these pages? If not, insist that training programs include more course content on risk factors, examples of problem situations, and techniques to overcome them.

2. Effective Prevention = Effective Prescreening. Objectively and dispassionately review the scope and depth of the history you are now routinely taking and the possibility that mailing it in advance might provide you with more time to observe and talk with the subject. How many of the complications in this book would you have missed because of deficiencies in your history forms and interview practices?

3. Effective Prevention = Adequate Informed Consent. How much of the informed consent procedures described in this chapter have you been using routinely? If you are not, your risk of precipitating complications, or being sued by a subject, increase with your percentile of noncompliance with the recommendations. If you doubt this, ask your lawyer or one familiar with tort law or malpractice cases.

4. Effective Prevention = Careful Observation. The safest way to drive a car on ice is to maintain a "feel of the road" and proceed slowly, constantly aware of the turn of the road and any obstacles that lie ahead.

In a crisis situation in hypnosis you are very much "on thin ice" in a similarly dangerous situation. Even in the most routine use of hypnosis, a researcher in what appears to be a superficial, harmless procedure, a therapist using hypnosis for habit control (smoking, overeating, drinking) the "beast from the dark lagoon" can be released from the recesses of the subject's mind. As the cases in this book demonstrate, this can occur without deliberate probing. Hypnotists should observe breathing rate, muscle tones, facial expression, eyelid flutter where there has been none before, fidgeting, tremoring, etc.

Generally, complications are more likely to occur when subject, hypnotist or external physical risk factors combine and converge, the more serious the factors are to the subject the greater the severity of subsequent after effects. Usual causative factors are life stressors, past or present, real or imagined. Often, though, the precise cause is not easily identified and becomes apparent only as the complication occurs, when the subject is already in crisis. This makes diagnostic skill an important requisite: "The utilization of hypnotic techniques for therapeutic purposes should be restricted to individuals who are qualified by background and training to fulfill all the necessary criteria that are required for a full diagnosis of the illness which is to be treated" (Rosen, 1962, p. 687).

Hypnotists should "know themselves," their bias by training and study, preconceived notions, overgeneralizations and oversimplifications, and their verbal and nonverbal idiosyncracies. Helpful verbal traits are a warm, conversational tone and voice volume, comfortable rate of speech, word and imagery choice all appropriate to the subject's rate of speech and thought. Nonverbal behaviors should project an attitude and image of acceptance, support and competence. Hypnotist and subject should be comfortably seated, at optimal physical distance, with minimal touching but as therapeutically (clinical) or professionally (research) appropriate, and with casual neither avoidant nor excessive eye contact.

Hypnotists should know as much as possible about their subjects. This factor alone is in the author's opinion the major cause of complications in stage and large group hypnosis. You will never have too much information on any subject but you can get into considerable legal and ethical trouble if you don't have enough. In court if it is established you "could" have known, this becomes "could and should have known" and can very readily end in "could have and should have and didn't" which is "inadequate standards of care" and "lack of care and skill" or negligence.

There are three recommendations for effective preventive practice and crisis intervention:

1. The best intervention **just matches the scope, depth and severity of the complication.** It is as the Chinese philosopher LaoTse said 2500 years ago "do as you would cook a small fish — be careful not to overdo it." To be most effective, the intervention occurs as the complication arises and matches it like horse and rider through its duration or "life." This first recommendation is **strategic,** a risk management concept.

2. The best intervention is **carefully and consciously selected** from the hypnotist's repertoire, **best suited to subject needs and hypnotist's expertise.** This is the **tactical** component. Interventions for hypnosis complications are similar to those in psychotherapy, and books and journals on this subject can provide hypnotists with much helpful information to broaden their armamentarium of techniques. The following are a few the author has found helpful in hypnosis crisis intervention:

- Use your voice as a multilevel instrument
  (word choice suited to the subject;
  emphasis, pauses, intentional repetition)
- "Give 'em slack" — let them tell/show you what's happening
  wisdom of the body; Zen concept — they know what they need)
- Touch judiciously
  (can be a deepening technique;
  can be distancing, a contaminant)
- Careful, gentle cross-sectional probes
  ("What are you feeling right now?"
  "Tell me more about that . . .")
- Persuade gently
  (carefully, consciously choose
  how firm and forceful to be)
- "Talk 'em down" if agitated
  (keyed to hypnotist's knowledge
  of subject psychodynamics)

Tactics involve a "judgment call" by the hypnotist. Again, LaoTse gave us good advice: "Proceed as one would walk across a frozen river, be careful of your own weight."

3. The third recommendation is **operational,** and with no irreverence intended, is the "ten commandments" of preventive practice:

(1) A calm, reassuring voice
   (the "hidden message" is "it's gonna be OK")

(2) Simple, understandable word usage
   understood at all levels of consciousness)

(3) Accepting, nonjudgmental attitude
   (avoid commanding, correcting, criticizing)

(4) Listen well for "trance needs"
   (listening with the Zen "third ear")

(5) Observe carefully and objectively
   (see what's there, not what you want to see,
   or are afraid to look for)

(6) Small, easy steps toward self control and security
   (like systematic desensitization or behavior modification)

(7) Slow, gradual pacing and tracking
   (be patient, use pauses;
   paced with subject's mental speed)

(8) Reassure with simple verbal cues or "pats"
   ("Yes . . . fine . . . I see . . . Uh-huh . . . good . . . ")

(9) Give permission and power
   (let subject choose what to remember,
   how fast/slow to think and feel)

(10) Always be aware of the subject's vulnerability
   (Subjects are in a strange world inside their minds;
   be there with them)

SUMMARY

Because it is becoming increasingly apparent that hypnosis can, directly or indirectly, contribute to or precipitate severe complications, even life threatening medical emergencies (Kleinhauz and Beran, 1981), there is a question of who should hypnotize. Medicine, dentistry and clinical and experimental psychology each have their own national board which awards a diplomate in hypnosis based on written, oral and practice examinations. These diplomates have more formal education and training than any other professionals using hypnosis and are ideal to be mentors and preceptors in the field. Where nations, states and provinces have legislation restricting the use of hypnosis, licensed physicians, dentists and clinical psychologists are most frequently those legally authorized to practice, provided they have training and experience in hypnosis in addition to their professional credentials.

There is a need to improve professional hypnosis training, which currently has limited coverage specific to hypnosis complications. This deficiency also exists in books and journals on the subject. Rising consumerism and medical malpractice litigation further emphasize the need for hypnotists to be more aware of risk management and preventive practices. Informed consent and documented standards of practice are steps toward a risk management system which would protect the public and help hypnotists ensure quality care to their subjects.

In addition to a risk management system, a strategy of delivered services, preventive practices and the tactics of practice, hypnotists need to continue to develop their expertise in identifying and treating complications. Knowing what to do and when to do it is a judgment call based on the knowledge, skills and abilities of the hypnotist. When a complication presents itself, the hypnotist must decide whether to let it continue and guide it, slow it down and phase it out, stop it abruptly, or escalate it and vent whatever is underlying. When dehypnotizing or rehypnotizing, the decision must be made whether to suggest the subject be amnesic to the trance experience, partial amnesia allowing the subject to decide extent of recall, or total recall. Each case must be decided individually.

Risk management is based on the legal and ethical concept that practitioners are accountable to consumers and to society for what they do. Professional licensure has traditionally been offered to the public as assurance of competence but it has been criticized also as a "trade guild" or "empire building ploy." Professional associations have set standards of practice, but in the case of hypnosis, lay practitioners have caused courts to take action, most recently to control the use of hypnosis in courtroom testimony. Unless there can be some agreed standard of practice to ensure that complications will be identified and treated with care and competence, courts and legislatures will do so without consulting us and enact restrictive laws and legal precedents.

Hypnotists interested in upgrading their skills, keeping abreast of research and development, and affiliating with colleagues to advance and protect the art and science of hypnosis can write to:

American Society of Clinical Hypnosis
Suite 336, 2250 East Devon Ave.
Des Plaines, IL 60018

Society for Clinical and Experimental Hypnosis
129-A Kings Row Drive
Liverpool, NY 13088
International Society of Hypnosis
111 North 49th Street
Box 144
Philadelphia, PA 19139

In 1984, the **Center for the Study of the Self** was founded, a private research center to study the nature of hypnosis and to serve as a clearinghouse for accumulating a database of hypnosis complications, their diagnosis, prevention and treatment. Appendix E is a complications report form developed at the Center. Hypnotists, concerned professionals and consumers are encouraged to report details of complications to help develop effective preventive practices:

Center for the Study of the Self
3804 Hawthorne Avenue
Richmond, VA 23222

CONCLUSIONS

The five chapters of this book lead to what will likely be unpopular and controversial conclusions. The data presented encompass the entire history of hypnosis, from earliest to most recent use, in clinical, research and stage entertainment settings. The consistent theme which emerges suggests the following:

Who is most competent in hypnosis? As in any scientific study or practice, the best educated, trained and experienced should teach, research and practice. For hypnosis, this is the multidisciplinary group of specialists with diplomates in hypnosis awarded by their respective national boards (medicine, dentistry, clinical or experimental psychology) and which require written, oral and practice examinations. These are the most qualified to inform the public about hypnosis and to teach colleagues in these major helping professions.

Regrettably, the practice of hypnosis should be regulated by law, restricted to licensed physicians, dentists and clinical or experimental psychologists who have received postgraduate doctoral-level training and supervised experience in hypnosis at the same level and standards of care as for their profession. Competence should be demonstrated by written, oral and practice examinations. These are the best qualified for the independent unsupervised level of practice, a role with which they are already familiar in their own professions, which bind them legally and ethically to strict standards. Anyone else using hypnosis would not be licensed and would require supervision by a licensed practitioner skilled in hypnosis. Social workers, nurses, Master's level counselors or psychologists, and postgraduate students in medicine, psychiatry, dentistry and psychology would require supervision.

A major revision of hypnosis training is long overdue and desperately needed, away from the "fast food" model of one or a few weekends and limited supervised experience toward a more gradual, systematic, closely supervised development of knowledge and skills over a longer time period. Clinical psychologists and psychiatrists require more training and supervised experience than dentists and physicians but anyone who will practice hypnotherapy needs more intensive training than is presently given in most programs. There is too little information on identification, prevention and treatment of complications. This should be an integral part of any training, included in written and oral examinations.

Finally, the customary ending "there is need for more research" is no pious platitude and is especially true in this area. We need a better system of identifying and reporting complications and it is hoped this book will stimulate further research and development in these areas.

To conclude this chapter and this book, the author searched for a quote that reflects the realities and responsibilities which must be considered if hypnosis is to be made and kept safe. The result of that search are these words of Thomas Huxley, published in 1868 in *A Liberal Education:*

> The chess board is the world, the pieces
>   are the phenomena of the universe.
> The rules of the game are what we
>   call the laws of Nature.
> The player on the other side
>   is hidden from us.
> We know that his play is
>   always fair, just and patient.
> But also we know, to our cost,
>   that he never overlooks a mistake
> Or makes the smallest allowance
>   for ignorance.

APPENDIX A-1

HYPNOSIS SCREENING REPORT	Name:_____ Today's date:_____ Address:_____ Zip:_____ Age:_____ Education (Yrs):_____ Phone:_____ _____ _____home_____office____
PHYSICAL HEALTH	Date last physical:_____ Findings:_____ Taking any medications?_____ If YES, which? Smoke?_____ If YES, how much? _____ packs a day Drink:_____ If YES how much and what? Drink coffee?_____ If YES how much? _____ cups a day
PREVIOUS TREATMENT	Ever referred for or received treatment for mental problems?_____ If YES, briefly describe: Still receiving treatment?_____ If so, by whom? Have any up-setting personal problems right now?_____ If YES, describe:
LIFE HISTORY	As an infant were you _____ sickly or _____ in good health? Describe childhood/teens: _____ happy _____ average _____ unhappy Parental relationships: _____ strict _____ average _____permissive Ever have or now have _____ fainting, _____ headaches _____ excessive dizziness (frequent) anxiety
FEARS OR PHOBIAS	List three situations or things which you find MOST FRIGHTENING: 1. 2. 3. Have nightmares or bad dreams?_____ If YES, briefly describe: Have any strange or unusual thoughts or ideas?_____
LEISURE ACTIVITIES	Hobbies: Favorite outdoor activities: Favorite movie: Favorite TV show:

APPENDIX A-2

HYPNOSIS PREP SHEET   Name:_____   **CONFIDENTIAL**

Instruction: Mark an X in the box which best matches your depth of involvement or level of interest in each of the following:

SUBJECT	LITTLE NONE	SOME	AV/ MOD.	MUCH	VERY MUCH
1. Previous hypnosis experience (I like it!)					
2. Positive attitude toward hypnosis (need it; want it)					
3. Emotional/psychological involvement in childhood crisis					
4. Emotional/psychological involvement in family problems					
5. I'd like to plan, think about, then go on vacation					
6. Absorption in hobbies, sports, and leisure activities					
7. I'd like to be an explorer					
8. I work hard and I play hard					
9. I'd like to fly a small plane around the world					
10. Interest, participation, belief, commitment to religion					
11. I'd like to fly in the Goodyear blimp					
12. I have close, sharing relationships with my friends					
13. I'd like to go back in history and see it firsthand					
14. I am a loyal and faithful follower					
15. Success in life demands self-discipline					
16. I like to kick off my shoes, relax, and take a catnap					
17. I feel that I have a strong personality and self-identity					
18. Sometimes I cry in the movies or watching TV					
19. I have/had a close relationship with my father (or mother or some other relative)					
20. I had a strict childhood					
21. I'd like to sail a schooner across the ocean					
22. I am concerned about personal problems					

APPENDIX A-2 *(continued)*

SUBJECT	LITTLE NONE	SOME	AV/ MOD.	MUCH	VERY MUCH
23. I'd like to be a kid again					
24. When I read something of interest I get deeply absorbed					
25. When deeply absorbed in something I lose track of time					
26. I'd like to climb a mountain					
27. I have a flair for the dramatic					
28. I'd like to be an astronaut and travel in outer space					
29. I like to play games of chance					
30. I like to daydream and fantasize					
31. I'd like to live in a foreign country					
32. It'd be fun to hang glide, skydive, race a car or boat					
33. I like to experience new things, taste new foods, etc.					
34. I like to listen to and "go" with music					
35. I like to travel					
36. I am deeply interested in my own behavior and mind					
37. I am curious, adventurous, and explorative					
38. I like to ride the mountain speedway					
39. When I make up my mind I am firm and don't change it					
40. I have a high degree of self-control and self-discipline					
41. I have a very high sense of duty and responsibility					
42. I am able to share deeply in intimate loving relationships					
43. I am able to regularly achieve sexual orgasm					
44. I feel I know myself pretty well					
45. I am more curious than fearful of death					
46. I'd like to "mellow out", relax deeply, and "let go"					
47. I'd like to downhill ski					
48. "Growing up" is a lifelong, continuing process					
49. Life is pretty much what you make of it					
50. I'd like to be a really good artist or sculptor					

APPENDIX B-1

MEDICAL-SOCIAL PROBLEMS
CHECKLIST

Name:_____ Today's date:_____

Address:_____ Home phone:_____

MEDICAL HISTORY	Family	You	Condition	SURGERY (list, give year):
	_____	_____	Cancer	
	_____	_____	Heart disease	
	_____	_____	Asthma, allergies	
	_____	_____	Diabetes	
	_____	_____	Headaches	ACCIDENTS (describe, give year):
	_____	_____	Seizures	
	_____	_____	Alcoholism	
	_____	_____	Suicide	Medications you are now taking:
	_____	_____	Other:	

CHECK ANY THAT APPLY *(you now have or have ever had)*

_____Allergies
_____Migraines
_____Dental problems
_____Sinusitis
_____Headaches
_____Insomnia
_____Bad dreams
_____Night sweats
_____Poor appetite
_____Nausea/vomiting
_____Diarrhea/constipation
_____Weight changes
_____Stomach pain
_____Bleeding
_____Frequent urination
_____Dribbling (urine)
_____Short of breath
_____Anxiety attacks
_____Pain in chest
_____Breathing trouble
_____Cough
_____Convulsions
_____The shakes
_____Fainting
_____Skin rash
_____Infections
_____Other:

HABITS   *How much?*
_____Smoking
_____Drinking
_____Overeating
_____Coffee/tea/cola
_____Drugs
_____Nervous mannerisms
_____Workaholic

SOCIAL PROBLEMS
_____Family
_____Marriage/partner
_____Job
_____Religion
_____School
_____Friends, neighbors
_____Children
_____Other:

SEXUAL PROBLEMS
_____PMS/menstrual
_____Impotence
_____No orgasm
_____Miscarriage
_____Other:

PROBLEMS NOT LISTED ABOVE, OR SPECIAL CONCERNS:

APPENDIX B-2

ANXIETIES, FEARS, AND
PHOBIAS DATASHEET

Name:_____ Today's date:_____
Address:_____ Zip:_____

SCARY PLACES Check only those you find especially scary. . .

	Accompanied	Alone
____Theaters, auditoriums	____	____
____Stores	____	____
____Classrooms	____	____
____Restaurants	____	____
____Elevators	____	____
____Garages	____	____
____High places	____	____
____Enclosed spaces	____	____
____Open spaces	____	____
____Buses	____	____
____Trains, subways	____	____
____Airplanes	____	____
____Boats, ships	____	____
____Expressways	____	____
____Parties	____	____
____Walking streets	____	____
____Standing in line	____	____
____At home alone	____	____
____Far from home	____	____
____Being watched	____	____
____Traffic	____	____
____Being lost	____	____
____Speaking in public	____	____
____Being criticized	____	____
____Being naked	____	____
____Being injured	____	____
____Sight of blood	____	____
____Going to the dentist	____	____
____Going to the doctor	____	____
____Surgery	____	____
____Needles/injections	____	____
____Funeral homes	____	____
____Dying (you!)	____	____
____Dying (others)	____	____
____Being trapped	____	____
____Water or drowning	____	____
____Snakes	____	____
____Bugs	____	____
____Other (list them:	____	____
_____	____	____
_____	____	____
_____	____	____

WHAT COULD HAPPEN? Check as many as you feel might happen to you in a crisis situation. . .

____throw up   ____pass out, faint
____brain tumor   ____heart attack
____choke and die   ____act silly, crazy
____go blind   ____lose control
____scream   ____hurt somebody
____paralyzed   ____wet your pants
____other: _____

PREVIOUS REACTIONS How you've reacted in the past in actual situations. . .

____palpitations   ____chest tightness
____numb arms, legs   ____fingers tingle
____short of breath   ____dizzy
____blurred vision   ____nausea
____lump in throat   ____knot in stomach
____dry mouth   ____rubber legs
____sweating   ____cold clammy
____confused   hands
____shaking   ____vomited
____other: ____derealization
                         (this can't be
                         happening!)

HOW YOU FELT How you usually feel in a scary situation. . .

____depressed   ____angry
____panic!   ____super scared!
____can't cope!   ____wet your pants
____can't breathe   ____run away
____confused   ____weird thoughts
____other:

INTERVIEW CHECKLIST
Earliest experience:_____
Prior treatment:_____
Attack frequency:_____
Pleasurable activities:

APPENDIX C-1

AGREEMENT AND CONSENT
TO TREATMENT

_____

I, _____ , agree to begin a program of treatment services to be provided to me by _____ as follows:

PROBLEM TO BE TREATED:

TREATMENT GOAL(S):

TYPE OF TREATMENT:

RESULTS SOUGHT:

RISKS:

FREQUENCY OF SESSIONS: **ESTIMATED NUMBER OF SESSIONS:**

EVALUATION PROCEDURE:

TERMINATE/TRANSFER: By my choice only (the client or patient).

COST:

_____ has explained the foregoing information and has advised me of alternatives to treatment by him such as psychiatrists, clinical social workers, nurse practitioners, counselors, the local mental health clinic, other psychologists, religious or pastoral counselors, self-help groups, or no treatment at all, and I have decided to proceed with treatment by him. I understand that all sessions are confidential but that _____ can divulge information with my written consent, by a judge's direct order, when my life or others are in clear and imminent danger, or when there has been child abuse. In agreement with these conditions, I sign my informed consent.

_____   _____
     Date

                    WITNESSED and agreed to _____

APPENDIX C-2

CODE OF ETHICS

(My pledge to you)

I RESPECT YOU as an individual,
    with dignity and integrity,
    morals, values and standards,
    a unique personality, and
I will do nothing to harm you.

YOU HAVE RIGHTS
    to accept or decline treatment,
    to stop treatment at any time,
    to ask questions and get answers.
I respect your rights.

OURS is an ethical, professional
    relationship,
    sensitive and confidential,
    without deception or deceit.
We will do only what we agree to do.

I HAVE AN OBLIGATION
    to know about you (take a history),
    to explain what we will do and
    how we will do it,
    to explain later what we've done,
    to take care of you
to protect your mental health.

APPENDIX D-1

HYPNOSIS DATA SHEET

Name: _____ Date:_____

Address: _____ Phone:_____

Age: _____ Educ. yrs: _____ Occupation: _____

Referred by: _____ Phone: _____

Address: _____

HISTORICAL DATA

CHILDHOOD
_____ Parenting
_____ Siblings
_____ Peers
_____ School
_____ Illnesses
_____ Traumas:

_____ Psychological:

TEENS-ADULT
_____ School
_____ Past jobs
_____ Present job
_____ Marriage
_____ Being parent
_____ Sexual
_____ Military
_____ Illnesses
_____ Traumas:

_____ Psychological:

CURRENT
_____ Conflicts
_____ Fears (name 3):
_____ 1.
_____ 2.
_____ 3.
_____ Illnesses
_____ Now an MH patient?
_____ Ever MH patient?
_____ Hobbies (risk):

HYPNO-AWARENESS
_____ Hypnotized before?
_____ Ever seen it done?
_____ Know about it?

PRE-INDUCTION INTERVIEW

_____ Rationale explained. Goals:
_____ Attitude; expectations
_____ Definition: "Feels like . . .
_____ Technique described
_____ All hypnosis is self-hypnosis
_____ Won't do anything against your will
_____ Won't touch (or will), ask questions
_____ Risk: Rare; any side effects will be treated
_____ OK to clear throat, scratch an itch
_____ Existential check (How're you feeling now?)

TEST INDUCTION

_____ eyelid flutter   _____ ireg. breathing
_____ anxiety   _____ movements
_____ tearing   _____ tremors
_____ facies   _____
IMPRESSION:

POST-INDUCTION SELF-REPORT

_____ heaviness   _____ lightness
_____ numbness   _____ tingling
_____ warm   _____ cold
_____ OOBE   _____ fantasy
_____ headache   _____ dizziness
_____ busy mind   _____ no effect whatever
IMPRESSION:

OBSERVATIONS

Case # _____ Name _____

Date this report _____ _____

                                              Hypnotist signature

APPENDIX D-2

HYPNOTHERAPY DATA SHEET	Name: _____ Age: _____
	Address: _____ Phone: _____

INDUCTION CHECKLIST	**SESSION**	**DATE**	**OBSERVATIONS/CONCLUSIONS**
_____ Position	1		
_____ Progressive relaxation	2		
	3		
_____ Countdown or press shoulder	4		
_____ Visual Imagery	5		
_____ ocean beach	6		
_____ forest scene			
_____	7		
_____ instructions: deep breath; countdown	8		
	9		
INDIVIDUALIZED TREATMENT	10		
_____ Posthypnotic suggestion: "Next time. . ."	11		
_____ Fix the seal	12		
_____ Wake 1-5	13		
TREATMENT OBJECTIVES	14		
Habit control	15		
_____ obesity			
_____ nicotine			
_____ alcohol			
Other(s):			

TREATMENT TERMINATION REPORT (Note progress, present condition)

Date termination and report _____   Hypnotherapist _____

©1981 Frank J. MacHovec, Ph.D.

APPENDIX E

HYPNOSIS COMPLICATIONS REPORT	Person making this report: _____ Date: _____ Address: _____ Phone: _____ Best day/time to call: _____
HYPNOTIZED PERSON DATA	Name: _____ Age: _____ Address: _____ Phone:_____ Subject's diagnosis (if any) at time complication occured Ever experience similar symptoms? ___ Yes ___ No Ever referred or treated for mental problems? ___ Yes ___ No Any known medical problems? ___ Yes ___ No Legal/corrections history if any: If taking medications, specify name, potency, dosage:
HYPNOTIST DATA	Hypnotist: _____ Phone: _____ Address: _____ Hypnotist's degree if known: _____ Licensed professional? _____ Hypnosis was used for:
COMPLICATIONS DESCRIPTION	Describe the complication (___ mental ___ physical ___ both): Onset: ___ during hypnosis___ during and after___ after hypnosis Intensity: ___ mild ___ moderate ___ severe Duration (specify number of minutes, hours, days or months): Treatment provided (when, what, by whom): Any lasting after effects?

REFERENCES

American Psychiatric Association (1982). *Desk reference to the diagnostic criteria from DSM-III*. Washington, DC: Author.

Applebaum, P. S. (1984). Hypnosis in the courtroom. *Hospital and Psychiatry, 35,* 657-658.

Argyris, C. (1968). Some unintended consequences of rigorous research. *Psychological Bulletin, 70,* 185-197.

Ault, R. L. (1979). FBI guidelines for use of hypnosis. *International Journal of Clinical and Experimental Hypnosis, 27,* 449-451.

Barber, T. X. (1961). Antisocial and criminal acts induced by hypnosis. *Archives of General Psychiatry, 5,* 301-312.

Baumann, F. (1986). Is it necessary to tell patients you are using "hypnosis"? In B. Zilbergeld, M. G. Edelestein & D. L. Araoz (Eds.), *Hypnosis questions and answers* (pp. 413-415). New York: Norton.

Biddle, W. E. (1967). *Hypnosis and psychosis.* Springfield, Illinois: Thomas.

Bjornstrom, F. (1970). Hypnotism: Its history and present development (1887). In M. M. Tinterow (Ed.), *Foundations of hypnosis, from Mesmer to Freud.* Springfield, Illinois: Thomas.

Boring, E. G. (1950). *A history of experimental psychology.* (2nd edition). New York: Appleton-Century-Crofts.

Bowers, M. K. (1956). Understanding the relationship between the hypnotist and his subject. In M. V. Kline (Ed.), *A scientific report on "The search for Bridey Murphy."* New York: Julian Press.

Bramwell, J. M. (1956). *Hypnotism: Its history, practice and theory (1903).* New York: Julian Press.

Cheek, D. B. & LeCron, L. M. (1968). *Clinical hypnotherapy.* New York: Grune and Stratton.

Coe, W. C., & Ryken, K. (1979). Hypnosis and its risk to human subjects. *American Psychologist, 34,* 673-681.

Conn, J. H. (1972). Is hypnosis really dangerous? *International Journal of Clinical and Experimental Hypnosis, 20,* 64-79.

Council on Scientific Affairs (1985). Scientific status of refreshing recollection by the use of hypnosis. *Journal of the American Medical Association, 253,* 1918-1923.

Diamond, B. L. (1983). Shirley: A correct decision. *American Psychological Association Division 30 Psychological Hypnosis Newsletter.* p. 3.

Dywan, J. D., Hamilton, E. P., & Orias, E. (1983). The use of hypnosis to enhance recall. *Science, 222,* 184-185.

Echterling, L. G., & Emmerling, D.A. (1984). Impact of stage hypnosis. *Submitted for publication.*

Edelstein, E. J., & Edelstein, L. (1945). *Asclepius.* Baltimore: Johns Hopkins University Press.

Erickson, M. H. (1962). Stage hypnosis back syndrome. *American Journal of Clinical Hypnosis, 3,* 141-142.

Erickson, M. H., & Rossi, E. L. (1981). *Experiencig hypnosis: Therapeutic approaches to altered states.* New York: Irvington.

Estabrooks, M. H. (1943). *Hypnotism.* New York: Dutton.

Evans, F. J. (1983). Forensic uses and abuses of hypnosis. *American Psychological Association Division 30 Psychological Hypnosis Newsletter.* pp. 2, 6.

Faw, V., Sellers, D. J., & Wilcox, W. W. (1968). Psychopathological effects of hypnosis. *International Journal of Clinical and Experimental Hypnosis, 16,* 26-37.

Freud, S. (1950). Psychopathology of hysteria. In J. Strachey (Trans.), *Collected papers of Sigmund Freud,* pp. 256, 262, 270. London, England: Hogarth Press.

Freud, S. (1953). My views on the part played by sexuality in the etiology of the neuroses. In J. Stachey (Trans.), *Collected papers of Sigmund Freud,* Volume VII, p. 27. London, England: Hogarth Press.

Fromm, E., & Shor, R. E. (1972). *Hypnosis: Research developments and perspectives* (pp. 459-463). Chicago: Aldine Atherton.

Fulgoni, D. (1983). Shirley: A bad decision. *American Psychological Association Division 30 Psychological Hypnosis Newsletter.* pp. 5, 6.

Gravitz, M. A., Mallet, J. E., Munyon, P., & Gerton, M. I. (1982). Ethical considerations in the professional applications of hypnosis. In Rosenbaum (Ed.), *Ethics and values in psychotherapy.* New York: Free Press.

Gross, S. (1985). *Of foxes and henhouses.* Westport, Conn.: Quorum Books.

Hare-Mustin, R. T., Marecek, J., Kaplan, A. G., & Liss-Levinson, N. (1979). Rights of clients, responsibilities of therapists. *American Psychologist, 34,* 3-16.

Hilgard, E. R., & Loftus, E. E. (1979). Effective interrogation of the eyewitness. *International Journal of Clinical and Experimental Hypnosis, 27,* 342-357.

Hilgard, J. R. (1974). Sequelae to hypnosis. *International Journal of Clinical and Experimental Hypnosis, 22,* 281-296.

Hilgard, J. R., Hilgard, E. R., & Newman, M. R. (1961). Sequelae to hypnotic induction with special reference to earlier chemical anesthesia. *Journal of Nervous and Mental Disorders, 133,* 461-478.

Janet, P. (1925). *Psychological healing* (Vol. 1). New York: MacMillan.

Kaufman, M. R. (1962). Historical background. In Group for the Advancement of Psychiatry, *Symposium 8, Medical uses of hypnosis,* pp. 647-652. New York: Author.

Kleinhauz, M., & Beran, B. (1981). Misuses of hypnosis: A medical emergency and its treatment. *International Journal of Clinical and Experimental Hypnosis, 29,* 148-161.

Kleinhauz, M., & Beran, B. (1984). Misuse of hypnosis: A factor in psychopathology. *American Journal of Clinical Hypnosis, 26,* 283-290.

Kleinhauz, M., Dreyfuss, D. A., Beran, B., Goldberg, T., & Azikri, D. (1979). Some after-effects of stage hypnosis. *International Journal of Clinical and Experimental Hypnosis, 27,* 219-226.

Kline, M. V. (1976). Dangerous aspects of the practice of hypnosis and the need for legislative regulation. *Clinical Psychologist, 29,* 3-6.

Kost, P. F. (1965). Dangers of hypnosis. *International Journal of Clinical and Experimental Hypnosis, 4,* 220-225.

Kroger, W. S. (1963). An analysis of valid and invalid objections to hypnotherapy. *American Journal of Clinical Hypnosis, 6,* 120-131.

Kroger, W. S. (1977). *Clinical and experimental hypnosis* (2nd ed.). Philadelphia: Lippincott.

Laurence, J., & Perry, C. (1983). Hypnotically created memory among highly hypnotizable subjects. *Science, 222,* 523-524.

Levitt, E. E., & Hershman, S. (1962). The clinical practice of hypnosis in the United States: A preliminary survey. *International Journal of Clinical and Experimental Hypnosis, 32,* 55-65.

Lidz, C. W., Meisel, A., Zerubavel, E., Carter, M. Sestak, R.M., & Roth, L. H. (1984). *Informed consent: A study of decision making in psychiatry.* New York: Guilford Press.

Loftus, E. E. (1979). *Eyewitness testimony.* Cambridge, MA: Harvard University Press.

MacHovec, F. J. (1975). Hypnosis before Mesmer. *American Journal of Clinical Hypnosis, 17,* 215-220.

MacHovec, F. J. (1979). The cult of Asklipios. *American Journal of Clinical Hypnosis, 22,* 85-90.

MacHovec, F. J. (1981). Shakespeare on hypnosis: The Tempest. *American Journal of Clinical Hypnosis, 24,* 73-78.

MacHovec, F. J. (1984, August). Hypnosis misuse (Letter to the editor). *APA Monitor,* p. 2.

MacHovec, F. J. (1985, August). Complications and unwanted side effects from the use of hypnosis and hypnosis-like states. *Presented at the 10th International Congress on Hypnosis and Psychosomatic Medicine, Toronto, Ontario.*

MacHovec, F. J. (in press). Public hypnosis and the public conscience. *Psychotherapy in Private Practice.*

Marcuse, F. L. (1959). *Hypnosis: Fact and Fiction.* Baltimore: Penguin.

Marcuse, F. L. (1964). *Hypnosis throughout the world* (pp. 81-235). Springfield, Illinois: Thomas.

Masling, J. (1966). Role-related behavior of the subject and psychologist and its effects upon psychological data. In D. Levine (Ed.), *Nebraska Symposium on Motivation.* Lincoln, Nebraska: University of Nebraska Press.

Meares, A. (1960). *A system of medical hypnosis.* New York: Julian Press.

Meares, A. (1961). An evaluation of the dangers of medical hypnosis. *American Journal of Clinical Hypnosis, 4,* 90-97.

Meldman, M. J. (1960). Personality decompensation after hypnosis symptom suppression. *Journal of the American Medical Association, 173,* 359-364.

Miller, M. M (1979). *Therapeutic hypnosis.* New York: Human Sciences Press.

Oppenheim, H. (1904). *Diseases of the nervous system,* p. 747. Philadelphia: Lippincott.

Orne, M. T. (1962a). On the social psychology of the psychological experiment with particular reference to demand characteristics and their implications. *American Psychologist, 17,* 776-783.

Orne, M. T. (1962b). Problems and research areas. In Group for the Advancement of Psychiatry, *Symposium 8, Medical Uses of Hypnosis,* pp. 676-689. New York: Author.

Orne, M. T. (1965). Undesireable effects of hypnosis: The determinants and management. *International Journal of Clinical and Experimental Hypnosis, 13,* 226-237.

Orne, M. T. (1971). The simulation of hypnosis: Why, how and what it means. *International Journal of Clinical and Experimental Hypnosis, 4,* 183-210.

Orne, M. T. (1972). Can a hypnotized subject be compelled to carry out otherwise unacceptable behavior? *International Journal of Clinical and Experimental Hypnosis, 20,* 104-117.

Orne, M. T. (1979). The use and misuse of hypnosis in court. *International Journal of Clinical and Experimental Hypnosis, 27,* 311-341.

Perry, C. (1977). Uncancelled hypnotic suggestions: The effect of hypnosis depth and hypnotic skill on posthypnotic persistence. *Journal of Abnormal Psychology, 86,* 570-574.

Putman, W. H. (1979). Hypnosis and distortions in eyewitness memory. *International Journal of Clinical and Experimental Hypnosis, 27,* 437-448.

Robinson, D. N. (1974). Harm, misuse and nuisance: Some first steps in the establishment of ethics for treatment. *American Psychologist, 29,* 233-238.

Rosen, H. (1953). *Hypnotherapy in clinical psychiatry.* New York: Julian Press.

Rosen, H. (1960). Hypnosis: Applications and misapplications. *Journal of the American Medical Association, 172,* 683-687.

Rosen, H. (1962). Technical modifications: therapeutic considerations. *In Group for the Advancement of Psychiatry, Symposium 8, Medical uses of hypnosis,* pp. 661-675. New York: Author.

Rosenthal, R. (1966). *Experimenter affects in behavioral research.* New York: Appleton-Century-Crofts, 1966.

Rosenthal, R. (1969). Interpersonal expectations: Effects of the experimenter's hypnothesis. In R. Rosenthal & R. L. Rosnow (Eds.), *Artifact in behavioral research.* New York: Academic Press.

Rosenzweig, S. (1933). The experimental situation as a psychological problem. *Psychological Review, 40,* 337-354.

Rostafinski, M. J., & MacHovec, F. J. (1986). *Variations in training programs in clinical hypnosis, U. S. and abroad.* Submitted for publication.

Sakata, K. I. (1968). Report on a case failure to dehypnotize and subsequent reputed after effects. *International Journal of Clinical and Experimental Hypnosis, 16,* 221-228.

Shevrin, H. (1972). The wish to cooperate and the temptation to submit: The hypnotized subject's dilemma. In E, Fromm & R. E. Shor (Eds.), *Hypnosis: Research developments and perspectives.* Chicago: Aldine-Atherton, pp. 527-536.

Shor, R. E., & Orne, M. T. (Eds.). (1965). *The nature of hypnosis.* New York: Holt, Rinehart & Winston.

Smith, M. C. (1983). Hypnotic memory enhancement of witness: does it work? *Psychological Bulletin, 94,* 387-407.

Spiegel, H., (1967). Is symptom removal dangerous? *American Journal of Psychiatry, 123,* 1277-1283.

Stam, H. J., & Spanos, N. P. (1982). The Asclepian dream healings and hypnosis: A critique. *International Journal of Clinical and Experimental Hypnosis, 30,* 9-22.

Teitelbaum, M. (1965). *Hypnosis induction technics* (pp. 104-112). Springfield, Illinois: Thomas.

Walker, S. (1967). *Psychiatric signs and symptoms due to medical problems.* Springfield, Illinois: Thomas.

Watkins, J. G. (1983). The baby and the bath water. *America Psychological Association Division 30 Psychological Hypnosis Newsletter.* pp. 4, 7.

Weizenhoffer, A. M. (1957). *General techniques of hypnotism.* New York: Grune and Stratton.

West, L. J., & Deckert, G. H. (1965). Dangers of hypnosis. *Journal of the American Medical Association, 192,* 9-12.

Williams, G. W. (1953). Difficulties in dehypnotizing. *Journal of Clinical and Experimental Hypnosis, 1,* 3-12.

Wolff, E. (1936). *Practical Hypnotism* (p. 25). New York: Max Holden.

Yarmey, A. D. (1979). *The psychology of eyewitness testimony.* New York: Free Press.

Zelig, M., & Beidleman, W. B. (1981). The investigative use of hypnosis: A word of caution. *International Journal of Clinical and Experimental Hypnosis, 29,* 401-412.

AUTHOR INDEX

SUBJECT INDEX

COMPLICATIONS INDEX